£17.50

The Environmental Risks from Biotechnology

The Environmental Risks from Biotechnology

by M. Chiara Mantegazzini

Prepared for the Commission of the European Communities, Directorate-General Environment, Consumer Protection and Nuclear Safety

Frances Pinter (Publishers), London
and Wolfeboro N.H.

© ECSC, EEC, EAEC, Brussels and Luxembourg, 1986

Publication No. EUR 10742 of the
Commission of the European Communities,
Directorate-General Telecommunications, Information
Industries and Innovation, Luxembourg

LEGAL NOTICE
Neither the Commission of the European Communities nor any
person acting on behalf of the Commission is responsible for
the use which might be made of the following information.

First published in Great Britain in 1986 by
Frances Pinter (Publishers) Limited
25 Floral Street, London WC2E 9DS

Published in the United States of America in 1986 by
Frances Pinter (Publishers) Limited
27 South Main Street, Wolfeboro, NH 03894-2069

All rights reserved. No part of this publication may be reproduced,
stored in a retrieval system, or transmitted in any form, by any means,
electronic, mechanical, photocopying, recording or otherwise,
without the prior permission of the publisher.

British Library Cataloguing in Publication Data
Mantegazzini, M.
    Environmental risks and biotechnology.
    1. Biotechnology industries—European
    Economic Community countries—Safety
    measures   2. Biotechnology industries—
    Hygienic aspects—European Economic
    Community countries
    I. Title   II. Commission of the European
    Communities. *Consumer Protection and*
    Nuclear Safety
    361.1'196696'094   HD7269.B4/
    ISBN 0-86187-658-X

Library of Congress Cataloging-in-Publication Data
Mantegazzini, M. Chiara, 1957–
    The Environmental Risks from Biotechnology.
    Bibliography: P.
    Includes index.
    1. Biotechnology—environmental aspects—
Europe.   2. European Communities.   I. Title.
TD 195.B58M36 1986   363.7'0094   86-22698
ISBN 0-86187-658-X

Typeset by Joshua Associates Limited, Oxford
Printed by Biddles of Guildford Limited

Table of Contents

1 Introduction and summary   1

2 Definition of biotechnology   10

3 The techniques   12

4 Fields of application   17

5 The impact on the environment   30

6 Potential risks from industrial applications   38

7 Potential risks from agricultural and environmental applications   59

8 Biotechnology regulation in the Member States of the European Community   82

9 Biotechnology regulation in the Community   99

10 The existing regulatory framework for protection of the environment   105

11 Product-specific regulations related to biotechnology   125

12 Conclusions and recommendations	128
Notes	129
Annexes	134
Index	161

1 Introduction

The Third Environmental Action Programme (O.J. C46/1–17 February 1983, Chapter II) outlines the need to develop an environmental policy orientated towards the prevention of pollution. In addition, it establishes a comprehensive framework to prevent and control damage from commercial chemical substances. It is expected that new developments in biotechnology will produce different problems because this technology derives from and relates to natural environmental and ecological systems in many ways. Two aspects are particularly important:

(1) biotechnology, to a great extent, can serve to protect the environment,
(2) due to its very nature, such technology could disturb natural ecosystems and the associated biological and nutrient cycles.

The purpose of this study is to evaluate risk to the environment of products and processes in the field of biotechnology. The study:

- defines biotechnology;
- identifies the possible fields of application and their potential for industrial development;
- examines the potential risks associated with biotechnological processes which may arise at different stages: research, production and processing, marketing of microbial cells or their products;

Introduction

- reviews relevant laws, regulations and guidelines already established or in preparation at national, Community and international levels;
- indicates suitability of existing legislation, identifies gaps in the legislative coverage of biotechnology and considers the need for changes in Community laws and regulations to protect human health and the environment.

SUMMARY

Commercial products of new techniques in biotechnology will be available in the very near future. Because of the many possible applications, biotechnology may offer significant benefits to society by alleviating problems of disease and pollution and increasing the supply of food, energy and raw materials. However, as with any new process, there are questions about the human health and environmental implications of developing and using organisms for commercial purposes. These arise particularly from:

- large-scale industrial production of commodities through biotechnologies, involving risks from accidental release of living organisms and other specific safety aspects;
- the use of new or novel organisms for various environmental and agricultural purposes.

POTENTIAL RISKS FROM INDUSTRIAL APPLICATIONS

1. Research

Although no major problems have so far arisen, registration of work and appropriate containment levels are generally required, because there is continued concern about the following:

(1) spontaneous mutations in pure and mixed cultures when growth conditions are changed;
(2) toxins produced in thermophilic systems;
(3) modification of viruses during fermentation and their impact on other organisms;
(4) the cloning of toxic genes and the introduction of antibiotic resistance genes into micro-organisms not known to acquire them naturally.

2. Industrial processing

Hazards associated with the new genetic techniques for industrial processing may be assessed in the same way as those associated with non-engineered organisms, that is, by examining the intrinsic hazardous properties of the components used in the engineering process. Industrial hazards are more likely to be quantitative than qualitative and possible hazards can affect (a) persons who work in industrial production units and/or (b) persons and environment exposed to emissions.

2.1 The working environment

Most of the studies on the survival of, and gene transfer by generically engineered bacteria have used mutatant debilitated strains of *Escherichia coli* K-12. Other debilitated genera and species of micro-organisms can be used as hosts for engineered genes, but less information is available about bacteria other than *E.coli*. To avoid the potential danger of transmitting antibiotic resistance traits to pathogenic micro-organisms, whenever the use of antibiotic resistant strains is necessary adequate physical and biological safety precautions should be adopted. Only resistance to antibiotics that are not applied therapeutically should be used for marking strains. Where new micro-organisms are used of which we have no experience, or which have undergone substantial alteration in structure, rapid ways must be found of establishing their potential as pathogens.

2.2 Persons and environment outside the industrial installation

There has been insufficient research to evaluate the mediating influence of environmental factors, both biotic and abiotic, on the survival, establishment and growth of, and genetic transfer by, genetically engineered bacteria. However, the limited data that are available indicate that abiotic factors—such as pH, salinity, aeration, water content—and biotic factors—such as competition between the engineered microbe and the indigenous microbiota of the specific habitat being studied, generation time and plasmid size —exert an influence on the survival and transfer of genetic information by engineered organisms under natural conditions.

Because the growth of dangerous human and animal pathogens must continue on a small scale for diagnostic, research and vaccine production purposes, a list of such organisms should be drawn up and their handling made subject to license in the EEC. (At present, most countries do not list pathogenic organisms and some do not regulate the handling of dangerous pathogens.) However, in general the technology of containment itself appears to offer no special problems once the standards are set.

It is also important to establish that those involved in the work are responsible for safe work practices and for protecting the environment from contamination. Education and training are equally important in minimizing risks. Appropriate education should ensure that managers and industrial workers at all levels are informed about potential hazards and control measures. Training is a prerequisite to ensure that appropriate safeguards associated with industrial processes can be completely carried out.

2.3 Products

The appearance of genotypic or phenotypic changes in the various stages of the production process can cause undesirable properties in the product. The prevention of these changes is in the interest of the manufacturer, to maintain maximum profitability. Close process control and quality control of the product should enable the manufacturer to recognize hazards related to changes in the culture media and producing micro-organisms and could be assured by adherence to Current Good Manufacturing Practices as adopted in the United States. Research is needed to develop more convenient and quicker methods to detect microbial contamination within large culture volumes and to discover the potential for phenotypic and genotypic changes in micro-organisms.

3. POTENTIAL RISKS FROM AGRICULTURAL AND ENVIRONMENTAL APPLICATIONS

The changes in established systems brought about by introduction of exotic organisms vary widely. Competition, predation, parasitism or pathogenicity on the part of the invader can result in reduction or exclusion of existing forms. These stresses may also contribute to the establishment of other exotics to cause more ecological imbalance. Over the long term, however, a new organism may finally become integrated into a system. This integration is usually marked by a significant reduction in numbers of individuals. Why this occurs may be obscure as are the reasons for their explosive spread in the first place.

A review of the literature reveals that firstly, the more disturbed, artificial, or simplified an environment is, the less likely it is that a new balance can establish in a reasonable time. Secondly, even if an exotic becomes integrated and is no longer explosive, the system it entered has been modified and is perhaps simpler.

Unfortunately, no quantitative assessment of the risk of ecological damage by genetically manipulated organisms can be made on the basis of existing ecological knowledge. Efforts directed at deriving numerical probability for establishment and disruption based on the number of negative results that have already occurred would not be meaningful or fruitful for a number of reasons. First, what constitutes a negative result is not always clear, and may be a matter of interpretation. Second, very little attention is every paid to exotic organisms unless they become a problem, and we do not really have any idea how many have been introduced and failed to get a foothold. Third, ecologists do not understand enough about the complex interactions in an ecosystem to be able to predict the outcome of an introduction. The many uncontrolled and unknown factors mean that new situations must be considered individually.

3.1 Micro-organisms

Direct or indirect changes may affect the environment when deliberate or accidental release of micro-organisms takes place. Many risks have been foreseen but they cannot, at present, be fully characterized because insufficient information is available. There is an acute need for further research to allow adequate assessment of the risks and the formulation of meaningful guidelines. In addition to producing genetically engineered microbes, the biotechnology industry should assess the survival, establishment and growth of, and the genetic transfer by, these engineered organisms under natural conditions.

3.2 Plants

The introduction of new plants into the environment must always be accompanied by careful evaluation and constant oversight by the agricultural research community. Previously, the continuing development of new crop cultivars created the need to assess their agricultural stability and also to assure the preservation of the environment. Undesirable properties have usually been detected during experimental trials.

A plant obtained through genetic engineering must be observed under its natural growing conditions to determine whether the genetic trait which was modified is expressed as expected or predicted and to be sure that it does not have any detrimental or debilitating effects.

Applications of biotechnology could be employed in agricultural programmes in less developed countries where, commonly, supplies of fertilizers and lime are scarce, the potential for irrigation is small, and adequate support for technological innovation is limited. In addition, marginal land in northern European countries could be reclaimed for forest products and biomass.

The potential impact of genetic technologies on conservation of natural ecosystems, genetic variability and crop vulnerability is still unknown. Because of the demands of genetic engineering, the need to collect, preserve and study diverse populations is now greater than ever before. Public funds should assist research projects that could entail environmental benefits, because these are of less commercial interest and unlikely to be undertaken by industry.

4. BIOTECHNOLOGY REGULATION IN THE MEMBER STATES OF THE EUROPEAN COMMUNITY

Member countries of the European Community have differing regulatory situations in accordance with their different levels of experience in the new techniques. This means that the level of expertise on evaluation procedures and the attitudes established in respect to environmental risks are correspondingly diverse.

- In the United Kingdom and in Denmark regulations have been set up for observance of notification and safety

guidelines in research work with recombinant DNA (r-DNA).
- In the Federal Republic of Germany notification is compulsory for research work supported by the government while it is voluntary for research funded privately.
- In the Netherlands, France, Greece and Ireland notification applies on a voluntary basis and reviews of protocols take place.
- In Belgium, notification is voluntary for medical research funded by the government but review of protocols is not conducted.
- In Italy and Luxembourg no specific recommendation applies to biotechnology research or its developments.

From the viewpoint of promoting international trade and a common market within the Community, harmonization of biotechnology regulations is necessary, to prevent any temptations for countries to 'under-cut' one another with less stringent regulations. At present, no specific provisions exist for the use in the environment of genetically engineered organisms.

5. THE EXISTING REGULATORY FRAMEWORK FOR PROTECTION OF THE ENVIRONMENT

The adaptability and suitability of existing environmental regulations with respect to biotechnology will probably be a matter of further research. However, environmental regulations, as they stand now, were not designed to control the risks which could arise from accidental or deliberate release into the environment of new living organisms. Further, even if some aspects of these existing measures could be made applicable, there would still be significant areas of concern where various open environment releases would not be subject to these procedures.

In some instances it may be possible to extend present legislation to cover various aspects of environmental concern. Nevertheless, the number of regulations to be amended is relevant and the inherent complexity and importance of biotechnology applications do not suggest that this piecemeal approach will provide complete, consistent and predictable regulation to ensure environmental protection and productive industrial growth.

2 Definition of biotechnology

In undertaking a study on any aspects of biotechnology the scope of the study must be defined. The definition of biotechnology, in fact, has been a matter of endless debate and many organizations and working parties have worked out definitions which range considerably according to the interest of those involved, but all have recognized the need for a common working definition to assess, for example, the extent of funding or the relevance of publications.

A report of Organization for Economic Cooperation and Development (OECD)[1] has proposed a definition of biotechnology covering a wide range of disciplines while excluding some agro-food activities: 'the application of scientific and engineering principles to the processing of materials by biological agents to provide goods and services'. A second relevant definition, from the United States Office of Technology Assessment (OTA) report[2] includes:'any technique that uses living organisms (or parts of organisms) to make or modify products, to improve plants or animals or to develop micro-organisms for specific uses'. As stated in a note of the Concertation Unit for Biotechnology in Europe (CUBE—Commission of the European Communities), the OTA definition is better than the OECD's but still far from perfect; it omits the provision of services and by its specificity in fact arbitrarily limits and excludes some possible fields of application (for example, detoxification).

For the purpose of this report, which will focus on the

evaluation of environmental risks of biotechnological products and processes and on the suitability of existing legislation, a sufficiently broad definition will be adopted, which will correspond to the scientific and practical reality of biotechnology as described in the Forecasting and Assessment of Science and Technology (FAST) report (Annex 1)[3] and to the definition proposed by CUBE (a streamlined version of OTA) as follows: 'the use or development of techniques using organisms (or parts of organisms) to provide or improve goods and services'. Other definitions are listed in Annex 2.

It should be stressed that biotechnology, far from being new, represents a developing and expanding technology based on a centuries-old foundation: people have deliberately selected organisms that improved agriculture, animal husbandry and food making since the beginning of civilization. More recently, a better understanding of genetics has led to more effective application of traditional genetics in such areas as antibiotic and chemical production. The rapid development of molecular biology and cellular biology in the last few decades has laid the scientific basis for entirely new technologies. Dynamic and progressive changes, some of which are illustrated in Annex 3, have characterized this sector ever since.

3 The techniques

Biotechnology is an integrated field of activity, relying on knowledge assembled and methods developed from many disciplines such as microbiology, biochemistry, physiology, enzymology, genetics and engineering. However, only a few techniques are going to be mentioned here because a comprehensive description would be beyond our scope.

Two major genetic discoveries, cell fusion and r-DNA techniques, by allowing the 'species barriers' to be overcome, greatly extend the horizons of biotechnology.

Genetic recombination is one way of increasing the diversity of organisms: it is the bringing together of genetic information to form new stable combinations or 'genotypes'. In nature, genetic recombination occurs between organisms of the same species or closely related ones.

Recombinant DNA technology allows direct manipulation of the genetic material of individual cells. The ability to direct which genes are used by cells, permits the development of micro-organisms that produce new products, existing products more efficiently, or large quantities of otherwise scarce products. This technology can also be used to develop organisms that themselves are useful, such as micro-organisms that degrade toxic wastes, or new strains of agriculturally important plants.

Cell fusion (or protoplast fusion), the artificial joining of cells, combines the desirable characteristics of different types of cells into one cell. This technique, unlike r-DNA, can be used when the characteristics of interest are controlled in a complex manner by a large number of genes so that large portions of genome must be combined. This technique, first described in 1975 by Kohler and Milstein, has been used to incorporate into one cell the traits for immortality and rapid proliferation, from a mouse skin cancer cell (myeloma), and the ability to produce useful antibodies, from a specialized cell of the immune system. This resulted in a hybrid cell (hybridoma) which grew in vitro and produced a pure specific antibody. Never before could such pure (monoclonal) antibodies be produced: we had had to rely on impure mixtures of antibodies of immune animal serum to provide immunological protection against disease.

Owing to their purity and high specificity, monoclonal antibodies (MAbs) have great potential in many areas, including diagnosis and treatment of disease, purification and sorting techniques, and tissue typing for transplants. Moreover, r-DNA and MAb technologies can complement each other, because r-DNA can lead to the production of new compounds and MAbs can aid in their identification and purification.

However, the translation of scientific achievements into working processes involves scaling up dimensions and a whole set of engineering, biochemical and economic problems that do not exist with laboratory experiments. Besides these well-known discoveries, which help to develop possible valuable products, biotechnology relies on the vital contribution of engineering, which for several reasons has received less attention than the biological sciences:

- engineering advances are slower;
- the cost of engineering research and development with biological materials is high, often requiring the production of large batches of material to test the performance of a single unit operation;
- major breakthroughs are infrequent.

Essentially, fermentation on a large scale involves keeping billions of micro-organisms in the best growing conditions, providing inexpensive carbon, nitrogen, mineral sources and sometimes an oxygen supply, to maintain their metabolism and for the formation of product. The reactions involved must be regulated to avoid enzymatic chaos and prevent overproduction of any one product so that a nutritional balance within the cells is maintained. Once having obtained a large volume of culture broth, either in single batches or in a continuous-flow process, the product must be recovered from the mixture of cellular components, nutrients, wastes and water; these stages, subsequent to the fermentation process, are called 'downstream processing'.

Recent improvements in techniques for immobilizing cells or enzymes (immobilized biocatalysts) and in bioreactor and biosensor design are helping to increase production and facilitate recovery of many substances, with important repercussions on the economics of the process.

Bioprocesses have been or are being used for the:

- production of cell matter (biomass) for consumption as food and feed (baker's yeast, mushrooms, algae, and various forms of single cell protein);
- extraction of particular cell components (enzymes, nucleic acids);
- manufacture of chemical products arising from the metabolic activity of particular cells. These products vary from primary metabolites (with a simple molecular structure like ethanol or lactic acid), to secondary metabolites (with a more complex structure like antibiotics or hormones);
- catalysis of specific, single substrate chemical reactions, for example, the conversion of glucose to fructose;
- simultaneous catalysis—usually employing a mixed population of micro-organisms—of a wide variety of physical and chemical changes on complex and frequently variable feeds (as in the waste treatment procedures);

- collection and concentration of minerals from dilute solution.

Bioprocesses may offer the following advantages over conventional chemical processes:

- milder reaction conditions (temperature, pressure and pH) with the result that equipment requirements may be much less demanding;
- use of renewable resources as raw materials (for example, plant biomass, sugar and starch) providing the energy required for synthesis and the carbon skeletons for organic chemical manufacture;
- less expensive or more readily available raw materials;
- the biocatalysts (such as cells, enzymes, and so on) are normally highly specific; a single isomer, rather than multiple forms of the desired product, may be obtained;
- multiple catalytic activities leading to the formation of complex molecular structures can be carried out in a single-step process;
- improved process efficiencies (for example, higher yields, reduced energy consumption);
- improved processes and new products developed through genetic engineering technologies.

However, there are also conceivable disadvantages of bioprocesses, some of which are:

- the generation of complex product mixtures requiring extensive and costly separation and purification procedures;
- problems arising from the relatively dilute aqueous environment in which bioprocesses function:
  (a) low reactant concentration, hence low reaction rate;
  (b) large volumes of high quality water are needed;
  (c) large volumes of diluted and high biological oxygen demand wastes are generated;
  (d) complex and frequently energy-intensive recovery

methods for removing small amounts of products from large volumes of water are required;
(e) large equipment capacities with high capital investment per unit product delivered are involved;
- the susceptibility of most bioprocess systems to contamination by foreign organisms and the need to contain the primary organism so as not to contaminate the surroundings;
- the inherent variability of biological processes due to such factors as genetic instability and raw material variability;
- for genetically engineered systems, the need to contain the organisms and sterilize the waste streams, an energy-intensive process.

4 Fields of application

The existing state of biotechnology-based industries and the range of their activities have been documented in a number of governmental publications, private reports, proprietary information surveys and reviews in scientific, trade and financial journals in the past few years. Detailed information will not be given here of all the wide range of present and potential applications of biotechnology potentially affecting any current industrial biological process or any process in which a biological catalyst can replace a chemical one. Annex 4 summarized the major biotechnological applications, classified according to the sector of activity and the volume/value basis. The following brief comments focus sector by sector on those applications more relevant to the environment.

1. HEALTH CARE

The new biological techiques were first applied in the health care industrial sector. According to most commentators, products in this sector will dominate biotechnological developments over the next decade, mainly because:[4]

- r-DNA and MAb technologies were first developed with public funds directed towards biomedical research;
- pharmaceutical companies have had years of experience

with biological production methods and this experience has enabled them to take advantage of the new technologies;
- pharmaceutical products are of high added value and can be priced to recover costs incurred during research and development, making the health care sector a good place to begin the costly process of developing a new technology.

Annex 5 lists some of the products that might be manufactured employing biotechnology developments in the future.

2. CHEMICALS

2.1 Bulk chemicals

Although fermentation processes have been used for centuries in the production of food and drugs, the chemical industry is largely petroleum based and orientated to the physical sciences and engineering.

Applications in the sector of bulk chemicals would include the production of feedstocks by fermentation of plant material (renewable resources) and wastes from other industries. Conventional fermentations often produce effluents of high organic content due to the incomplete utilization of substrates, for example, spent grain and stillage in brewing, distilling or alcohol production and whey from the milk-processing industries.

Of the one hundred organic chemicals of greatest significance for chemical synthesis, only six have been produced in commercial quantities by fermentation: acetic acid, acetone, n-butanol, ethanol, glycerol and isopropanol. For economic reasons, all are currently produced by chemical synthesis.

It has been estimated[5] that biotechnological feedstocks and processes for the production of bulk organic chemicals could not easily nor fully displace existing petrochemical feedstocks and production processes within the next ten years. However, ethylalcohol produced by bioprocesses is already cost competitive with that obtained from ethylene and it can be said that, with the increase of oil prices and supply problems, these technologies would become more attractive.

2.2 Specialty chemicals

The above estimates do not necessarily take into account the expansion of uses for existing organic chemicals (for example, oxychemicals and their derivatives) or the development of new, 'nontraditional' commodity organic chemicals such as biopolymers.

According to the OTA[6] report, the production of specialty high added-value chemicals represents one of the largest opportunities for biotechnology because of the diversity of potential applications (see Annex 6: potential chemical products using r-DNA technology).

Projected initial impacts will be most likely in the production of natural metabolites such as aminoacids, peptides, enzymes, vitamins and organic acids: increase in sales volume could shift these high-price, low-volume chemicals to low-cost, high-volume commodity chemicals.

Moreover, genetic engineering opens possibilities of creating new biomaterials: 'Protein engineering' does not only involve moving genes around but includes building new ones from scratch and using living organisms to produce materials that never existed in nature. It is believed, for example, that very strong fibres or plastics could be manufactured by making specific modifications in genes to improve the products they can make, such as spider webs or wool, which could be made noncombustible. Demands for more specialist materials are increasing in areas as diverse as aerospace and health care.

Enzymes, catalysts of natural biochemical processes which are easily destroyed in non-physiological conditions, might be made temperature- or acid-resistant so to improve their performance in industrial manufacturing processes.

2.3 Oils and fats

At present, oils and fats constitute only about 2 per cent of the raw materials for synthetic organic chemicals and are concentrated in industries such as paints, surface coatings and detergent products. Recent work has demonstrated a variety of new uses for oils and fats, such as using linseed oil for coating polymers or soyabean oil for nylons.

Furthermore, vegetable oil-based plasticisers can be made from a number of vegetable oils and these could become very useful because dioctyl phthalate, the current most commonly used plasticiser, is rapidly coming to be regarded as an undesirable environmental contaminant. Vegetable oil-based plasticisers should have better biodegradability and lower toxicity.

The use of genetic manipulation, in addition to the existing methods of plant and animal breeding, to modify the fatty acid composition of oils and fats, is a field of active industrial research.[7] Soyabean and rapeseed are being genetically modified to produce high-priced specialty oils that are now derived principally from coconut and palm kernel. The added-value is clear, for example, the cost of oil used in the manufacture of detergents will be reduced by half when produced from a genetically engineered commodity crop.

3. BIOELECTRONICS

Biotechnology could be used to develop improved biosensors or new conducting devices called biochips. Biosensors, using enzymes and MAbs, would improve the sensitivity and rapidity of control methods of industrial bioprocesses; biochips, using proteins as semiconductors, may also have great potential applications. Some advantages would include small size, reliability and potential for self assembly. However, these are one of the more distant applications of biotechnology.

4. FOOD PROCESSING

Some present practices and future developments in the food and beverage industries are highlighted in Annex 7. Future developments will be in the production of enzymes, flavourings and additives.

4.1 Enzymes are commercially among the fastest growing group of food additives. At a rough estimate, more than fifty enzymes are used in the food industry in brewing, cheese making, in the production of fruit juices and wine, in starch and sugar processing and as meat tenderizers.

Traditionally, enzymes have been extracted from plants and animals. Rennin, for instance, the enzyme used to curdle milk to make cheese, was extracted from the stomach of unweaned calves, but now, like many other enzymes, it is produced from engineered micro-organisms.

The FAO/WHO Expert Committee on Food Additives is concerned about the uncontrolled use of food enzymes. It argues that these enzyme preparations could contain potentially harmful by-products and contaminants because of the great variability of the production processes and the mutations which may occur in micro-organisms. Its view is that chemical and microbiological specification should be required as well as extensive toxicological testing for enzymes derived from less well-known microbes (for example, pullulanase, which is an enzyme derived from a pathogenic bacterium to be produced by harmless *E.coli*).

4.2 Myco-protein, now being produced for full-scale testing in the United Kingdom, is regarded as the basis of a new food technology and was formally sanctioned as safe for human consumption by the United Kingdom government in 1980. A fungus is grown on glucose syrup produced from starch converting the carbohydrate to first-class protein. The micro-fungus, unmodified by genetic engineering, has a thread-like shape, conferring valuable textural properties on the product. It can be shaped to simulate animal protein textures and flavour: meats, poultry and fish have all been imitated satisfactorily. The advantage is the very high rate of food conversion of myco-protein, about 1:1, compared with 10:1 or worse for animal husbandry. Also, the organism has a doubling time of only 5 hours.

4.3 Proteins are being developed as 'functional' food ingredients. For example, processes are being developed by I.C.I. to fractionate the 'Pruteen' biomass into protein isolates, nucleic acids and phospholipids, and the potential role of the protein isolates is currently being investigated. Food manufacturers are becoming more aware of the function of protein ingredients in food and companies can selectively modify the protein they produce—for example, by limited enzymic hydrolysis. The use of specific enzymes or alterations in the process conditions may soon lead to tailored production of proteins for specific applications.

5. AGRICULTURE

5.1 Animal husbandry

(1) The management of the reproductive biology of animals for agriculture, including aquaculture, will provide several applications as 'wild animal farming'.

One example[8] is the adaptation of sheep to harsh ecosystems, such as dry or hilly regions. New genotypes produced by the introduction of the necessary genes from hardier wild strains into domestic more productive strains achieve a high efficiency in converting rangeland forage to meat. However, the development of these applications ultimately depends on advances in embryo culture and embryo transfer techniques.

(2) Several products for animal health including vaccines, for example, for foot and mouth disease, rabies and bovine papilloma, diagnostics and therapy products, which will find commercial markets similar to those predicted for human health applications of monoclonal antibodies.[9]

(3) The availability of animal growth hormones from genetically engineered micro-organisms is likely to have a significant economic effect in the future. Recombinant bovine growth hormone (BGH) has been long-term tested in the United States and is now waiting for regulatory approval. Bovine growth hormone improves both milk and beef production: cows injected with BGH for 188 days give 41 per cent more milk, and health or milk composition were unaffected. For dairy cows, daily injection is expected to be economical.[10]

Other developments in this sector are the large-scale production of single cell protein, 50,000 tonne 'Pruteen' fermenter in the United Kingdom, and the significant improvements in the design of anaerobic fermentation processes to produce methane (biogas) on site from animal wastes.

5.2 Pest control

Approximately 1,500 naturally occurring micro-organisms or microbial by-products have been identified as potentially useful insecticides. Because of its multiple advantages—selectivity, absence of chemical residues and much slower development of resistance—bio-control of pests offers an interesting alternative to current practices and a challenge for applied molecular biology and ecology. Although several bio-agents have met the safety, efficacy and environmental criteria established by regulations, *Bacillus thurigiensis* sp. is the only one that has been developed successfully as a commercial insecticide on a very large scale.

Research goals for genetic engineering are to increase the spectrum of pests controlled by biological pesticides, to increase pesticide efficiency as toxic agents, and to improve production processes and economics (at present *Bacillus thurigiensis* can cost 20–25 per cent more than comparable chemicals).

If a microbe's ability to survive and replicate is to be increased or if its range of activity is to be broadened greatly, the possible environmental side effects must be carefully assessed before field-testing and large-scale release.

5.3 Plant cultivation

A major goal of research has been to develop new varieties of plants into marketable seeds or vegetable propagules through the introduction into plants of genes such as those controlling resistance to low temperature, high salt concentration, herbicides or specific disease (viral, fungal and bacterial), and in the long-term, the cloning of the nitrogen-fixing capability. Another possibility is that of raising the nutrient value and the efficiency of growth of strategic crops, such as corn. The time frame in which research developments will be sufficiently successful to have a massive impact on plant cultivation is difficult to ascertain.

Existing systems for plant production may at first be subject to only minor modifications, but as research increases the understanding of organized genomes, it is not inconceivable that food, fibre and energy sources will be produced by entirely new processes.[11]

So far, clonal propagation has been applied to potato plants for the propagation of special, virus-free plants and to improve large-scale propagation and mechanical delivery systems. Somaclonal variation technology has developed new tomato varieties with added value for the consumer and the processors. Fresh market tomato varieties supply the consumer with tomatoes of uniform size, good texture, red colour, good flavour and good shelf life making tomatoes available on a year-round basis. Tomatoes developed for the processor focus on increased solids, flavour, colour, sugars, acidity, and so on.

Micro-organisms can express some useful properties which could be used in crop production:

- *Pseudomonas syringae*, either chemically muted or genetically engineered, to protect crops from mild freezing temperatures;
- *Rhizobium*, the nitrogen-fixing bacteria, is being thoroughly investigated with the aim of transferring the ability to fix atmospheric nitrogen to agronomically important species other than legumes.

6. ENERGY

(1) Applications of biotechnology to energy production are well known, particularly in areas concerned with agriculture, such as biomass production and conversion.

The Brazil 'alcohol' programme based on sugar cane and the United States 'gasohol' programme have provoked a debate concerning the desirability of using arable land for the production of fuel rather than food and possible ecological disturbances. It is generally agreed that biomass production in regions like Europe, where sugarcane is not available, will not significantly contribute to biofuel production, at least until there are major improvements in hydrolysis for lignocellulose. An interesting chain of events has been mooted for the altered land usage for gasohol production in the United States in which increased corn production leads to increased prices for soyabeans and cotton, leads to synthetic fibres becoming competitive with cotton, which ultimately leads to increased consumption of petrochemicals to satisfy the increased fibre demand.[12]

On a local basis, biogas produced from anaerobic digestion of sewage and agricultural wastes can be a valuable energy source.

(2) New techniques are being developed to enhance oil recovery such as:

- the microbial production of compounds to be added to reservoirs (bio-polymers increase water viscosity so that it has greater power to displace crude oil and biosurfactants are emulsifiers and decrease oil viscosity);
- the injection of micro-organisms into the reservoirs that will produce surfactants and polymers or carbon dioxide.

One particular bacterial polysaccharide has been considered for injection into oil wells where it should make the oil more easy to extract, but the high temperature found in many oil wells is a challenge as is also the fact that the micro-organism in question (*Xanthomonas* spp.) happens to be a plant pathogen.

Encouraging field tests results have been obtained with the injection of *Bacillus* and *Clostridium* species along with fermentable raw materials and mineral nutrients. This produced a copious amount of carbon dioxide, methane and some nitrogen in oil reservoirs. The carbon dioxide made the crude oil less viscous and the other gases helped to repressure the reservoir.

(3) The treatment of domestic and industrial liquid wastes is the basis for a well-developed biotechnology industry, initially aimed at the control of pollution but now advanced

in the production of feed and biogas from organic wastes. Some valuable wastes are listed in Annex 8. It should also be noted that biotechnological industrial processes can generate pollution problems: a large brewery produces 10,000 cubic metres per day of wastes, equivalent to the sewage load from a community of 200,000 people.

7. METAL EXTRACTION

Micro-organisms can be used to drain dissolved metals from mine workings, waste dumps and coal heaps. Microbial leaching is developed to extract copper and uranium from sulphide ores.

The technique exploits bacteria of the genus *Thiobacillus* (for example, *Thiobacillus ferrooxidans*), which can grow on carbon dioxide from the air and on energy released by oxidizing the sulphur found in a variety of metallic sulphides, thus producing metal ions and sulphuric acid. These bacteria live in acidic environments and in the presence of oxygen. Microbial leaching of ores could solve several problems: compared to the traditional extraction system (smelting) it is inexpensive, uses little energy and reduces pollution problems due to the production of sulphur dioxide, which is involved in acid rain. Moreover, extraction of low-grade ores becomes possible.

However, very little is known about the biochemistry and genetics of the organisms used: new approaches include studies of the mechanism of resistance to high metal concentration, in some cases plasmid borne, for the production of high-concentration resistant engineered organisms.

8. METAL RECOVERY

Micro-organisms can accumulate metal ions by absorbing them into the cell surface or by taking them up inside the cell. These mainly electrostatic processes are fast, reversible, and useful even if the cells are quiescent or dead. They can have overlapping uses: to recover valuable metals from industrial wastes, and to clean them up by removing toxic metals. Iron, copper, mercury, silver, cadmium, cobalt, nickel and zinc are some examples of the metals concerned.

Common objectives for new developments are the following:[13]

- to be comparatively small scale;
- to focus on extraction and recovery of precious metals;
- to solve more than one problem of cost, energy conservation or pollution at once;
- to offer some unique advantage rather than competing directly with conventional processes.

In fact, established biological extraction processes for copper and uranium have been widely applied because of the low market prices of these commodities; on the contrary high added-value products, like precious metals, could produce substantial rewards for research and development.

9. POLLUTION CONTROL

One of the major benefits to society which could derive from biotechnology is the improvement of environmental quality through the development of procedures to treat soil or water contaminated with pollutants.

Pollution problems can be divided into two categories: those that have been present for a long time in the biosphere—for example, most hydrocarbons encountered in the petroleum industry and human and animal wastes—and those of human origin—for example, pesticides. Chemicals of both sorts, sometimes appear in places where they are potentially or actually hazardous to human health or the environment.

Pollution can be controlled by micro-organisms in two ways: by enhancing the growth and activity of micro-organisms already present or near the site of the pollution problem, and by adding more (sometimes new) micro-organisms to the pollution site.

There is disagreement about the value of adding micro-organisms to decontaminate soil or water. One point of view argues that serious spills frequently sterilize soils, and that adding micro-organisms is necessary for biodegradation. The other contends that encouraging indigenous micro-organisms is more likely to succeed because they are adapted to the spill environment. Added bacteria have a difficult time competing with the existing microbial flora. In the case of marine spills, bacteria, yeasts and fungi already present in the water participate in degradation, no-one has been able to demonstrate the usefulness of added micro-organisms.

Genetics has apparently been little applied to pollution control: by far the best-known work is that of Dr. A. Chakrabarty who engineered two strains of *Pseudomonas*, each of which has the ability to degrade the four classes of chemicals found in oil spills. This was in 1972 and, while the original organisms have never been used in the open environment, Chakrabarty has since been working on bacterial strains able to break down more toxic industrial wastes. He obtained bacteria which could live on acetic acid (2,4,5-Trichlorophenoxy), one of the components of Agent Orange, as a staple source of carbon.[14]

The potential of genetic techniques is clear; the constraints are questions of liability in the event of health, economic or environmental damage, the view that added organisms are not likely to produce a significant improvement and the assumption that selling micro-organisms rather than products or processes is not likely to be profitable.

5 The impact on the environment

Ten years ago molecular biologists and biochemists imposed on themselves a moratorium on experiments involving the cloning of DNA molecules in host systems such as *E.coli*. The basic concern was that an organism might be created that could cause disease if it escaped from the laboratory. In response, involved scientists developed guidelines designed to prevent such organisms from getting out of the laboratory and to guarantee, if they did get out, that they could not survive. Guidelines developed for small-scale experiments enabled research to be carried out under safe conditions, showing that risks could be controlled in the laboratory. As the safety of r-DNA research itself was demonstrated, emphasis on the commercial aspects of biotechnology increased.

1. GENERAL CONSIDERATIONS

At the time of writing, commercial products of new techniques of biotechnology will be available in the very near future. Because of the many possible applications, biotechnology may offer significant benefits to society by alleviating problems of disease and pollution and increasing the supply of food, energy and raw materials. However, as with any new process, there are questions about the human health and environmental implications of development and applications on a commercial scale. These arise particularly from:

- large-scale industrial production of commodities through biotechnologies, involving risks from accidental release of living organisms and specific safety aspects;
- the use of new or novel organisms for various environmental and agricultural purposes.

(1) Natural environments have evolved a variety of checks and balances that hold the many species and populations in our surroundings in dynamic equilibrium. The various mechanisms that are responsible for the balance among species in natural environments hold in check the many pests and disease-bearing organisms that these environments contain. It is these very interactions that prevent most organisms from one habitat to becoming established in another. These same mechanisms probably will destroy most organisms that are introduced, just as they have eliminated most organisms that are transported from one environment to another. Because of these natural checks and balances, ecological or environmental upsets associated with newly arrived or rare organisms are uncommon.

However, exotic organisms do become established in ecosystems—both with and without harmful effects. A semi-quantitative analysis of the introduction of foreign organisms[15] noted that in 854 documented cases of exotic introduction, 71 of the cases resulted in the extinction of a natural population. It is important to note that the total number of cases (including those in which the novel organism did not survive) is not known, but estimates range among several thousands. The adverse environmental or public health impacts which could result from the accidental or deliberate introduction of a new or novel organism have been summarized as falling into three general categories:[16]

(1) ecological disruption due to lack of natural enemies;

(2) infectivity, pathogenicity or toxicity to non-target organisms (plants, animals, humans) and
(3) exchange of genetic material with other organisms or disruption of ecosystems.

Six components of risks have been identified.[17] The probability of each component occurring ranges from zero to one hundred per cent. These are (L = likelihood of occurrence).

L_1 incorporation of gene for hazardous trait into an organism (in the case of new organisms);
L_2 the chance of release into a natural environment;
L_3 the possibility that an organism will survive there;
L_4 the likelihood that it will multiply in that environment;
L_5 the possibility that it will make contact with a receptive environment (gene exchange; dissemination);
L_6 the chance that it will be harmful.

Harmful biotic environmental effects will occur as the result of a series of these events, with each event having its own particular chance of occurring. The likelihood of harm is equal to the product of the probabilities:

$$L = L_1 \times L_2 \times L_3 \times L_4 \times L_5 \times L_6$$

Therefore, if the likelihood of occurrence of any step is zero, the final outcome will be zero, or no harm.

If the new organism is deliberately released to perform some useful function and it does what is expected of it, only the last two factors need to be considered, that is, the contact with a receptive environment and the possibility of its causing harm. Such changes may also be the unanticipated results of the growth of the organism as it performs its designated functions.

Indeed, large uncertainties exist in anticipating the consequences of the introduction of new and novel organisms into the environment because of major gaps in knowledge.

The evaluation of some of the issues, like survival or multiplication, could be relatively simple: absence of information is a reflection of the lack of attention given to the problem, either by researchers or by regulators.

Attention has been given to the possibility of gene exchange in nature. Laboratory tests on micro-organisms, for example, suggest that genetic information may be exchanged in soil or water, but this information is limited and usually comes from studies that were conducted under highly artificial conditions.

Information about the dissemination of organisms from one site to another comes from monitoring the spread of human, animal and plant disease. It shows that certain micro-organisms and plants are transported enormous distances, that the spread of disease, the dispersal of pollen and the transport of micro-organisms can be documented. However, little can be predicted about dissemination because few of the elements which determine susceptibility to dispersal are known.

Uncertainty also exists about the potential for harm. The information on ecological effects of diseases of plants, animals and humans is abundant. Enormous numbers of human deaths have resulted from the introduction of micro-organisms into regions where the people were not previously exposed to the harmful agent; agricultural crops have also often been devastated following the introduction of a new disease agent and many of the disease-producing micro-organisms were genetically very similar to species that previously had little effect on the farmer's crop.

It is true that for introduced genetic material or organisms probabilities of survival, multiplication, gene transfer, dispersal and detrimental effects are quite small and therefore the probability of the final event in the sequence is even smaller. The potential environmental risks associated with the introduction of new or novel organisms into an ecosystem are best described as 'low probability, high consequence risks', that is, while there is only a small possibility that damage could occur, the damage that could occur is great.

Indeed, for a new technology characterized by a low probability of causing harm, the early stages of development need cause little worry. But, as the technology becomes more widely used and it moves in new directions, the likelihood of undesirable consequences increases. In other words, if an undesirable event has a probability of occurring once in 1,000 uses of a given technology, the risk from a few uses of that technology would undeniably be low. There is no room for complacency, however, if 600 or 1,000 or more uses are envisaged.

The chemical industry can be taken as an example. When only a few chemicals were in daily use their production was low and their sites of entry into the human environment were few, little or no hazard existed for society at large and no threat was posed to major ecosystems. The issues of human health risk and ecological perturbation became only too evident, however, once the annual production of some chemicals had exceeded hundreds of tons per year and the sites from which exposure could occur had become multitudinous.

(2) An important issue which has been intensely debated is whether a genetically engineered organism released into a new ecosystem would produce significantly different results from those caused by translocation of natural organisms.

On one hand, it has been stated that

the likelihood of an organism becoming established in a given environment is an ecological question regardless of the origin or nature of the differences that make such an organism new or novel. The proportion of the genome which is 'new' is not necessarily correlated with the degree of impact an organism can have on an ecological system. Analogy between recombinant organisms and introduced species is a valid one.[18]

Thus, although highly modified organisms can be produced, the interaction between the recombinant organism and an ecosystem may be similar to that which occurs for natural organisms because the impact of an organism is more a function of the relationship between its characteristics and its surroundings than of its origin.

On the other hand, it has also been said that the impact of an organism on the environment could depend on the technique by which it was produced. For example, if an entire gene has been removed by deletion an engineered organism could not revert to its original type. Thus, no possibility exists that unexpressed genes carrying potentially adverse characteristics would be present in the organism. In contrast, other techniques (such as protoplast fusion, ultra-violet or chemical treatment) involve the large-scale alteration of existing DNA or the addition of large amounts of foreign DNA. In this case some changes and additions could be caused to the new genome.

Characteristics	Whole organism Engineering	Cellular Engineering	Molecular Engineering
Level	Organism	Cell	Molecule
Processes	Breeding selection	Cell fusion Cell culture/ Regeneration	Recombinant DNA
Control over genetic change	Random	Semi-random	Directed
Primary genetic	Unknown	Semi-known	Known
Other genetic changes	Unknown	Unknown	Known
Number of variants needed	Large	Intermediate	Small
Species restrictions	Mainly within	Within and across	Within and across

Source: *Issues in Science and Technology* Vol. 1, No. 3, 1985, National Academy of Science.

Since amino-acid sequences of proteins determine their conformations translating the one-dimensional genetic message into the three dimensions, it is expected that small changes in amino-acid sequences should produce only limited changes in protein structure and progressively greater divergences in sequence should produce correspondingly greater divergences in structure. However, observation perhaps surprisingly shows the following:[19]

- very different sequences can, if they have arisen under selective constraints on function, still generate very similar structures;
- very similar sequences can, in some cases, generate very different structures, which makes it difficult to predict the structural perturbation caused by even a small change in sequence. However, small changes in structure that make far-reaching changes in functions have never been identified; most likely, selection will explore the 'immediate vicinity' of a function.

Moreover, it is important to realize that the selection of organisms for further use is based on the presence of particular attributes sought by the researcher. No attempt is made to screen for other attributes of the organism unrelated to the particular manipulated gene. As a result, the potential overall effect of the modified organism on its surroundings is not given sufficient consideration.

Unanticipated characteristics can be produced when two nonpathogenic micro-organisms are combined. In 1977,[20] two scientists in New Zealand, K. L. Giles and H. C. H. Whitehead, attempted to enhance the nitrogen-fixing capability of a certain species of pine tree by genetically modifying a fungus that inhabited the trees. In seeking to modify this fungus the scientists combined two normally non-pathogenic micro-organisms. This combination normally should have produced a harmless fused organism capable of bringing about the desired result. In contrast, however, one strain of the newly created recombinant fungus was pathogenic and another strain killed tree seedlings to which it was applied. While the effect of the modified organism was clearly demonstrated on the seedlings, it is equally possible that the pathogenic nature of the new organism could have gone unnoticed for many years. There are disease-producing organisms, such as the chestnut blight, that affect only mature trees and have no effect on seedlings. Consequently, a disease of this type would not have been detected by the procedures used in this experiment.

This experiment is not cited to indicate that all genetically transformed organisms are, or can be, dangerous. Many single gene alterations have no effect on the environmental impact of an organism, and typically such changes result in mutations that are lethal only to the organism itself. The experiment is cited simply as a real example of potential danger resulting from release of a genetically engineered organism. It demonstrates why a thorough examination must be made of the potential environmental effects of an organism before its large-scale release occurs.

6 Potential risks from industrial applications

Concerns raised by industrial applications of biotechnology involve hazards that might arise at different levels:

(1) risks at the laboratory and pilot plant stage;
(2) risks associated with industrial processing and by-product handling;
(3) risks associated with the products obtained.

Many considerations will be applicable both to the research and to the industrial processing stages. Some remarks will apply to all biotechnology applications. However, the risks associated with industrial production will not be examined here in detail for all fields of application. Indeed, many different kinds of products can be obtained biotechnologically and different problems may arise in relation to each area of production. Specific considerations, applicable to different kinds of products, will affect the accurate assessment of the risks arising from individual products and processes.

1. RESEARCH

In the early 1970s, when the technology of genetic manipulation was first developed, attention focused on the possible harm which could arise from r-DNA research work. This risk was associated with the occurrence of a possible series of events: the inadvertent incorporation of hazardous genes into micro-organisms, the escape from the laboratory into the environment, and the occurrence of ecological disruption. Codes of practice were prescribed in many countries while in some others no restrictions at all were imposed. The United States[21] and the United Kingdom[22] adopted the most stringent guidelines, which attempted to classify all possible experiments that could be done using r-DNA and assigned to each one a level of potential risk.

To accommodate these classes of hazard the guidelines defined corresponding levels of containment, specifying both the requisite features of laboratory design and the standards for laboratory practice. The guidelines also established local and central administrative structures to oversee compliance, to approve certain experiments and to modify the guidelines. Gradually, most experiments were downgraded to lower containment levels and certain experiments using common r-DNA vectors and host organisms were judged to be so safe that they were exempted from the provisions of the guidelines. Several lines of scientific evidence as well as the accident-free record of the research led to a widespread feeling that r-DNA techniques were safer than originally feared.

In the United States, an increasing number of requests were made for exemption from the initial prohibition against experiments involving volumes more than 10 litres. The great majority of these demands[23] originated from industry. However, as in most other cases, the American guidelines were mandatory only for institutions receiving public grants, and even for these institutions, the penalty for violating the guidelines was simply loss of funding. Over the years, large-scale procedures utilizing well-characterized host-vector systems were reviewed at local level only. In fact, it was accepted that the potential environmental problems associated with the use of well-characterized organisms in large-scale r-DNA production processes were similar to those associated with non-recombinant DNA large-scale fermentations, traditionally performed with good safety records.

1.1 Conclusions

Although no major problems have so far arisen, registration of work and appropriate containment levels should be required.

In fact, concerns remain with respect to:

(1) the spontaneous mutations in pure and mixed cultures when growth conditions are changed;
(2) toxins produced in thermophilic systems;
(3) modification of viruses during fermentation and their impact on other organisms;
(4) the cloning of toxic genes and the introduction of antibiotic resistance genes into micro-organisms not known to acquire them naturally.[24]

2. INDUSTRIAL PROCESSING

Today the large-scale industrial applications of new biotechnologies are raising questions of concern similar to those debated in the 1970s for research: containment problems are being discussed in respect to the change in scale taking place in biotechnology applications.

The expanding industrial applications of biotechnology, in fact, present some new features:

(1) the very large scale on which new micro-organisms are about to be grown for the first time; and
(2) the great number of small and medium enterprises from sectors which do not have the good safety record of traditional fermentation industries who will use biotechnology to produce existing and also completely new products.

The probability for inadvertent release occurring, which was low for operations on research scale, would be considerably increased with fermenters of large-scale capacity; also, security measures appropriate for volumes of thousands of litres cannot be simply extrapolated from guidelines developed for no more than 20-litre volume batches. Moreover, industrial voluntary compliance to guidelines may not be sufficient to ensure risk prevention and a positive public attitude.

The hazards associated with the new genetic techniques for industrial processing may be assessed in the same way as those associated with non-engineered organisms, that is, by examining the intrinsic hazardous properties of the components used in the process. However, though the hazards may not vary in quality, they do in quantity. Possible hazards can affect (a) persons who work in industrial production units or (b) persons and the environment exposed to its emissions (see Commission reports listed in notes 25 and 26 for a long discussion of the hazards).

2.1 The working environment

Micro-organisms are associated with people, animals and plants, their influence being, in most cases, beneficial. Some, however, can cause an antigenic response or disease in higher forms of life by multiplying within the body of the host, sometimes with the concurrent generation of toxins. Other micro-organisms produce toxic agents while outside the body which can subsequently cause disease when encountered by man and animals.

Those primarily at risk are the workers in the biotechnology factory. Here dense aerosols of micro-organisms are generated which, in the absence of suitable precautions, will be released into the factory atmosphere during recovery of products of fermentation by centrifugation, spray or flash drying, packaging and subsequent handling of the product, and the release of culture supernatant.[25]

The disease or the allergic response will depend on the virulence of the strain of the micro-organism, the dose (concentration and duration of exposure), and point of entry, and the immune status and state of health of the subject. For infection, the dose is generally the most important factor since, if it is large, natural immunity can be overcome. Indeed, if the dose is sufficiently great, organisms not normally pathogenic can infect people; they can also cause disease if they have access to usually inaccessible parts of the body, for example, the brain or spinal cord following an accident or surgery.[26]

There is no doubt that certain people develop allergies against a large number of naturally occurring substances and also against certain micro-organisms and their metabolic products. As soon as allergies appear in a biotechnological process, the allergic person should at once be removed from the biotechnological operation. Such allergies are particularly to be anticipated when micro-organisms exist outside culture media. It is advisable in such cases to determine whether the risk of exposure should be reduced for other persons, too, for instance, by avoidance of dust formation, wearing protective clothing, and so on.

There is a wide range of experience of the problems of working with enzymes from micro-organisms in biotechnological plants, particularly those producing specific enzymes (such as proteases) in large quantities. No danger for personnel is to be anticipated so long as the enzymes exist inside the cells. Hazards only appear when the enzymes are produced outside cells and thus can reach the personnel by means of the liquids or aerosols in which they exist.[27]

Very few pathogenic micro-organisms and viruses are likely to be used by industry. At present they are employed in the manufacture of vaccines or diagnostic reagents. Examples include *Bordetella pertussis*, *Mycobacterium tuberculosis*, and the virus of foot and mouth disease. The scale of operation is usually relatively small, with culture vessels of a few cubic metres capacity. Safety is ensured by physical containment of the organisms during cultivation, followed by inactivation before use. However, the cultivation of attentuated pathogenic strains must be monitored closely to detect back mutation to virulence, as is normal practice in vaccine production process.

Moreover, concerns have arisen on the use of E.coli in recombinant DNA research. Escherichia coli is a normal intestinal inhabitant of human beings, as well as of all warm-blooded animals. It has been suggested that genetically modified E.coli could colonize the body of the researcher and thus escape from the laboratory into the environment.[28] Furthermore, E.coli is 'sexually promiscuous' and can transfer genetic information, especially that on sub-cellular plasmids, for example, antibiotic resistance used as 'marker', to representatives of over 40 Gram-negative bacterial genera, including promoters of transmittable disease. Escherichia coli is also an 'abnormal' occupant of rivers, streams, lakes and estuarine waters and of soil in urban and agricultural areas, thereby extending the possibility of transfer to other bacteria in these environments.[29]

Because of the possibility of genetically engineered bacteria inadvertently escaping and subsequently becoming established in the environment, debilitated strains of E.coli can be used as the host cells for engineered genes. For example, the host E.coli K-12 strain ×1776 is sensitive to bile salts, thereby diminishing its potential survival in the gastrointestinal tract; has a generation time that is two to four times greater than that of wild type E.coli, thereby decreasing its ability to compete successfully with the indigenous microbiota and requires diaminopimelic acid, an amino acid not commonly present freely in nature, and thymidine, the absence of which causes cell death and degradation of DNA.

Also, presumably safe plasmids (called pBR322 and

pBR325), poorly mobilizable, can be hosted by the debilitated strains of *E.coli*. Several 'in vivo' studies have established that *E.coli* does not colonize the human intestinal tract, even after ingestion of billions of organisms by volunteers.[30] For example, elevated levels of this bacterium occurred in the faeces of human volunteers that had ingested strains of *E.coli* K-12 only on day 1 after ingestion, but thereafter the numbers rapidly decreased, and no K-12 strains were detected by days 3 to 5. Apparently, the indigenous intestinal microbiota has a selective advantage over introduced strains so that risk of colonization of the human intestine by *E.coli* K-12 appears to be very low.

Most of the studies on the survival of, and gene transfer by, genetically engineered bacteria have commonly used *E.coli* K-12, with emphasis on the debilitated strains of this host. Presumably also other debilitated genera and species of micro-organisms will be used as hosts for engineered genes, but little information is available on the survival and gene exchange, either in culture or under natural conditions of bacteria other than *E.coli*.

To avoid the potential danger of transmitting antibiotic resistance traits to pathogenic micro-organisms, whenever the use of antibiotic resistant strains is necessary adequate physical and biological safety precautions should be adopted. Only resistance to antibiotics that are *not* applied therapeutically should be used for 'marking' of strains. Where new micro-organism are used about of which there is no experience, or which have undergone substantial alteration in structure, ways should be found to allow possible infectivity to be quickly assessed.

2.2 Persons and environment outside the industrial installation

The unique hazards of biotechnological processes involve the release of viable micro-organisms to the environment.

The most significant point is that organisms are taken from the closely controlled environment of the fermenter, moved into more open conditions of the recovery plant, and then leave the industrial plant either as products or as wastes.

2.2.1 Processing

While it is unlikely that micro-organisms responsible for a major human disease will ever be grown on a very large scale, the increasing use of plant pathogens is cause for concern. Well-documented examples of catastrophes which periodically overtake agriculture show that fungal and bacterial plant disease can be of overwhelming economic importance.[31] Two examples illustrate some relevant aspects.

Tobacco blue mould. In 1958 *Peronospora tabacina* was imported to the United Kingdom under licence from the British Plant Health Authorities, for use in fungicide experiments. In that same year the mould appeared on tobacco plants grown for virus research at four research institutes (three in England, one in the Netherlands) and in a commercial tobacco crop in England. The following year the disease appeared in the tobacco fields of Belgium and the Netherlands. Thereafter, inoculum was so plentiful that the mould was able to spread all over Europe, the disease advancing in Germany at a speed of 5–20 km per week. After more than 400 years of cultivation in the absence of the pathogen, it is not surprising that the crop was so susceptible. Subsequently, by natural selection and careful breeding, resistance to the mould was increased to a point where the disease was no longer completely destructive and tobacco crops could again be grown.

Peronospora tabacina arose from a point source, but the fact that the source was a licensed laboratory emphasized the risks involved in handling, even knowingly, a non-indigenous pathogen capable of infecting an indigenous crop.

Yellow rust of the wheat cultivar Heines VII. In 1950 the wheat cultivar (cultivated variety) Heines VII was introduced into the Netherlands partly because it was resistant to all known races of yellow rust, the fungus *Puccinia striiformis f.sp.tritici*. Three years later a small area of crop was found to have been attacked by what appeared to be a new race of the rust. By 1956, following the spread of spores in the wind, Heines VII growing 800 km from the original source was affected. This is just one of many examples of new races of plant pathogens appearing in response to the development and widespread use of cultivars bred for resistance to known pathogenic races.

Plant diseases are ever present, they can sometimes become of great economic importance and are constantly evolving in response to the introduction of new crops and new cultivars. Also, it is well recognized that many fungal and some bacterial diseases can be spread by airborne transport of spores. Also some bacteria produce protective gums which allow them to survive in aerosols and to maintain infectivity. Those gums are produced industrially for applications in the field of fine chemicals and food processing.[32]

It follows that the release of plant pathogenic microorganisms as an aerosol, at times when atmospheric conditions favour survival and dissemination of the aerosol to a susceptible crop, constitutes a potential threat to plant health.

However, the selection pressures within a fermenter are different from natural ones and unlikely to encourage pathogenicity because no selective advantage would be conferred. No spread of plant disease has been attributed, up to now, to the industrial use of plant pathogens. However, the possibility of such events occurring may increase as many more phytopathogens, or micro-organisms derived from them, are grown on a large scale and others, specifically designed to degrade plant materials such as lignocellulose, might turn out to be pathogenic for some plants.

A previous Commission study[33] extensively reviewed present and possible uses of plant pathogens and discussed possible hazards deriving from their industrial production. It concluded that

if adequate precautions are employed, there is no reason to expect that the industrial processing of plant pathogens on a scale not exceeding a few 1,000 litres will lead to their release, either as bulk liquid or, much more importantly, as aerosol. But as the scale increases, so does the cost and the difficulty of containment for large volumes of water-saturated air. Therefore, a plant pathogen which is still fully virulent after growth in a fermenter cannot, at present, be handled safely on a very large scale; at least, in the vicinity of crops of suitable host plants. Clearly, a non-plant pathogen should be used, whenever possible, but if its use is unavoidable previous tests should be made of its pathogenicity when grown under the proposed conditions. If it is still pathogenic, further assessment involved in its use should be made because even at scales not exceeding a few thousand litres, the possible consequence of a system failure should still be considered.

2.2.2 *Accidental release*

Accidental release might, for example, follow mechanical failure, filter failure or 'foam out' of the culture. It could be caused by the fire and explosion of material such as methane, methanol, ethanol, acetone or butanol or explosion of dry powders. It could follow a natural event (earthquake, volcanic eruption, hurricane) a terrorist sabotage or an accident in a neighbouring plant (for highly industrialized areas). Waiting for an accident to arise would not be a wise strategy.

In the case of organisms which could cause harm to plants and animals in the environment, environmental consequences of accidental release would in addition to the quantity released depend on:

(1) the position of the factory in relation to susceptible environments;

(2) the meteorological and environmental conditions (dispersal, dilution, survival and ability to enter a host depend on the weather and the physiological state of the host);
(3) the strain and the physiological state of the micro-organism (pathogenicity, infectivity, spore production and aerostability depend on growth conditions in the fermenter).

2.2.3 Wastes

In many biotechnological processes, it is the reaction product that is of interest and not the cell itself. The cell is then filtered or centrifuged off in the first stage of further processing. Depending on the micro-organism used, it has to be determined whether the cell mass must be killed off before disposal, this can be done by heat sterilization or by addition of disinfectant. With pathogenic micro-organisms, not only the process effluent but also the cleaning water, even that used outside the equipment, should be collected through a closed pipework system and then rendered harmless either by thermal or chemical means. The disposal of viable non-pathogenic micro-organisms may interfere with effluent treatment processes or produce undesirable effects if discharged through a river or sea outfall: it will provide food and affect the flora of those waters by increasing microbial, algal and plant growth or supply competition, which might change the natural balance with unforeseeable results. Moreover, the continuing discharge to a stream of an extraction effluent containing a low level of antibiotic could create conditions ideal for the selection of organisms resistant to that antibiotic, with very undesirable consequences. The genetic information, if carried on a plasmid, could then be passed by stages throughout the microbial world. *Escherichia coli* and *Bacillus subtilis* are not normally pathogenic, but are potentially dangerous because they can acquire resistance to antibiotics which is then transmitted by plasmids to pathogens.

It is consequently desirable to reduce waste viability (biological activity), through chemical treatment or heat, before disposal. Incineration can also be used but is relatively costly because of the high moisture content of biomass. Volumes of spent broth are large (4,500 cubic metres/day from a 50,000 tonnes per annum single cell protein plant) and in sewage treatment terms they represent strong wastes with a high biological oxygen demand, but are generally free of specific highly toxic constituents and/or heavy metals, and their pH is near neutrality.

2.2.4 Containment

The problem which must be faced when a manufacturer wishes to introduce a new biotechnological process is to determine whether the organism on which it is based is capable of causing disease, and if it is, to decide the appropriate method of containment. In this process, not only should the health and safety of those directly involved in the work be considered, but also any possible effects on plants, animals or the environment.

A number of organizations, including the Netherlands Microbiological Society, the German Health Authorities, the World Health Organization, the United States Department of Health and Human Services and the United Kingdom Department of Health and Social Security and Health and Safety Executives have introduced classifications of micro-organisms into risk classes, usually 1–4 in increasing order of pathogenicity and on the basis of hazard to workers and general public. Uncertainties about the risks involved (pathogenicity may depend on dose and on conditions of fermentation) is reflected in the lack of complete unanimity on these classifications.

In an attempt to harmonize the differing descriptions, the European Federation of Biotechnology[34] proposed a classification into four groups (1–4) (see Annex 9) with a fifth one, Group E, containing those micro-organisms offering risks only to the environment, especially to animals and plants.

While each class 1–4 would be correlated to four agreed levels of containment, it was suggested that the appropriate level for micro-organisms in class E, should be evaluated case by case.

2.3 Conclusions

There has been insufficient research to evaluate the influence of environmental factors, both biotic and abiotic, on the survival, establishment and growth of, and genetic transfer by, genetically engineered bacteria. However, the limited data that are available[35] indicate that abiotic factors—such as pH, salinity, aeration, water content—and biotic factors—such as competition between the engineered microbe and the indigenous microbiota of the specific habitat being studied, generation time and plasmid size—are significant.

Because the growth of dangerous human and animal pathogens must continue on a small scale for diagnostic, research and vaccine production purposes, a list of such organisms should be drawn up and their handling, except under licence, forbidden in the European Community. (At present, most countries do not list pathogenic organisms and some do not regulate the handling of dangerous pathogens.)

In all other cases, suggested containment levels can vary from nothing more than is needed in good housekeeping, to complete isolation (that is, all outgoing materials, including such items as operators' clothing, kept for detoxification, all operations in enclosed buildings maintained at below atmospheric pressure with 'absolute' filtration of outgoing air, all personnel bathing and changing clothing on leaving). Because the cost escalates rapidly as the level of control increases it would be extremely important to achieve an accurate assessment of the risks so that precautions can be set at the right level. In fact, the technology of containment itself appears to offer no special problems once the standards are set. Established methods are available even for materials as potentially dangerous as materials of biological warfare.[36]

It is also important to appreciate that those involved in the work are responsible for safe work practices and for protecting the environment from contamination. Education and training are equally important in minimizing risks. Appropriate education should ensure that managers and industrial workers at all levels are informed about potential hazards and control measures. Training is a prerequisite to ensure that appropriate safeguards associated with industrial processes can be completely carried out.

3. PRODUCTS

In all fermentation processes the main danger is of undetected changes which could cause the product to have different characteristics than those expected. A biotechnology process will remain stable only as long as the micro-organisms on which it is based remain unchanged. Changes occur as a result of the high adaptability of micro-organisms to cope with changing environmental conditions. The response to a change is determined by genes: usually these remain unchanged and the response is due to metabolic systems activating some enzymes and inhibiting others, thus producing what is known as phenotypic variation. Sometimes, however, genetic change is by spontaneous random gene modification (mutation) or by acquisition from other organisms of genetic information, such as that contained in plasmids which confer resistance to some antibiotics and toxic heavy metal ions: this is known as genotypic variation.

3.1 Genotypic variation

There is little opportunity for new genotypes to become established in batch production since only relatively few generations of micro-organisms happen in each batch. However, in continuous fermentations, the content of the fermenter must be maintained genetically homogeneous for hundreds of days and millions of generations. With natural mutation rates as high as 1 in 10^7, and with the additional chance of contamination from outside the system, the potential for genetic change is considerably greater in continuous fermentations than in batch production. Only few of the genetic changes that occur give rise to organisms which can survive in the conditions set for the fermentation; only if the changes confer some specific advantage, do the organisms bearing them become established.

3.2 Phenotypic variation

Unlike mutation, phenotypic change is a response which is maintained only as long as the new environmental conditions persist. It should, therefore, be prevented by the close process control which will be used by the manufacturer to maintain maximum profitability of his process. A phenotypic variation can be induced, amongst others, by:[37]

(1) a change of growth method, for example, from batch to continuous culture;
(2) a change in growth rate which could be brought about by a change in medium supply rate, and aimed, for example, to increase production rate;
(3) a change of growth limitation which could be caused by an incorrect formulation of the medium or a partial failure or blockage in the supply of a component of the medium;
(4) a change of temperature which could result from a failure in the pH control system;
(5) a change of trace elements which could stimulate or inhibit the production of a particular metabolite, for example, the inhibition by iron of diphteria toxin production.

The European Commission has recommended that research effort should focus on more convenient and quick methods to detect microbial contamination within large culture volumes and discovering potential for phenotypic change and spontaneous mutation in micro-organisms.[38]

Products from biotechnology are so diverse and have so little in common that it is very difficult to make a valid generalization about them. Many are used or consumed in daily life, in, for example, food or health care products, and are already subject to a large number of laws and regulations, traditionally based on their intended use. For some others, such as novel food proteins, no specific regulation exists.

3.3 Novel food protein

In many member countries of the European Community, if novel food proteins are regarded legally as food, they are not subject to special legislative approval except for compliance with the general hygiene and labelling regulations. They can, therefore, be freely marketed. If they were to be defined as food additives, they would be subject to all the regulations applied to establish the safety of food additives. Indeed safety problems posed by additives and novel food proteins are somehow different and required tests for additives would not always be applicable. It is not clear when or indeed whether novel proteins should be defined as food additives for the purposes of regulation.

The term novel food protein applies to food proteins prepared by processes and techniques, including new or severely modified processes not previously used in food production; these

- concentrate the protein component of existing traditional food;
- modify the physical properties of protein rich food;

- allow the use of sources of protein not previously exploited for human consumption or consumed only in small amounts.

It includes novel enzymes from micro-organisms modified by genetic engineering. In addition novel proteinaceous materials used in animal feed may modify in a subtle way traditional food proteins and must, therefore, be considered for the safety and control of novel food proteins.

There are two clearly separable approaches to developing novel protein components and food. One approach modifies conventional food sources by a new process, the other exploits previously unused sources on a large scale for human food. Examples of the first group of novel food proteins are fish protein concentrate, texturized forms of soya products and proteins derived from bones and cheese whey. Examples of the second group are protein concentrates prepared from grass and leaves of non-food plants, selected micro-organisms and new species of fungi grown on substrates providing a cheap source of organic carbon or on waste products, and material from algae and genetically engineered organisms. Different safety considerations apply to each type of product.

To establish a basis for regulating consumer exposure and for protecting public health certain information is essential. The safety considerations involved in the evaluation of the suitability of novel proteins have two aspects.[39] The first aspect is concerned with known toxic or anti-nutritional factors which may occur in the raw material from which the novel food protein is prepared. Examples are biotoxins occurring in tropical fish, the gossypol pigment of cottonseeds and the mycotoxins produced by fungal contaminants of seed meals. The second aspect relates to possible toxic effects following the lifetime ingestion of unusual constituents of the novel food protein such as non-food amino-acids, cell wall constituents of micro-organisms, unknown mycotoxins, unusual fatty acids and carbohydrates, and toxic contaminants such as certain heavy metals.

No significant problems are likely to arise when texturized or otherwise processed proteins from traditional vegetable food are being assessed toxicologically. If the texturing or processing uses completely new procedures, then toxicological examination is necessary. Proteins derived from animal sources are likely to give rise to difficulties in toxicological assessment only if natural toxins or contaminants are present or if the processing introduces any chemical modifications.[40] Proteins derived from vegetable sources not previously used as human food require thorough toxicological testing, and single cell proteins must be tested additionally for microbial contamination by the producing micro-organism, for absence of pathogenicity of the source organism, and for adequate reduction in their nucleic acid content.

Additional toxicological problems may arise through the processing applied to the novel food protein to convert it into a form acceptable for incorporation into a human food. Furthermore, proteins derived from micro-organisms may contain purines, pyrimidines and amino-acids not normally found in the animal or human diet. Bacterial and fungal metabolites may contaminate the protein or may arise as a result of mutation in the source organism. Unusual fatty constituents, amino-acids and substrate residues may appear in the meat, eggs and milk of animals fed these materials. These products may also be contaminated by pathogenic microbes. The nutritional and toxicological consequences of ingesting these foreign materials will require investigation and assessment.

Guidelines have been prepared by several national and international authorities describing the toxicological studies considered adequate for assessing the health hazards of novel food including novel food protein (EEC Council, 1983; ACINF, 1984; IUPAC, 1984; PAG/UNU, 1983). Within the framework of Codex Alimentarius, the

Codex Committee on Vegetable Proteins is also in the process of elaborating standards for certain novel proteins and guidelines for their use.

Toxicity testing of novel food proteins involves special problems not encountered in the testing of food additives.[41] Novel proteins and particularly novel food, are frequently complex mixtures of substances likely to be consumed at high dietary levels and should, therefore, be regarded as food and not as food additives. Yet, several of the guidelines are based on the assumption that novel proteins should be tested in a similar way to food additives. If the levels of dietary incorporation exceed 1 per cent, the usual 100-fold safety factor applied in testing protocols for food additives cannot be achieved, because the highest dosage level to be employed in the feeding test cannot, in these circumstances, be at least 100 times the level of the human diet. In any case, it is not possible to add excessive amounts of certain non-toxic foods to the diet of test animals without affecting their health.

The major components of novel protein food are unlikely to be toxic. To observe any effects due to minor constituents requires long-term feeding at dietary levels so high that non-specific effects due to nutritional imbalance may confuse the results. For these reasons studies extending over the lifespan of the laboratory animals are of little value in assessing safety. The same argument applies to carcinogenicity because the tumour incidence in the test species can be altered by nutritional imbalance resulting from the excessive protein load, for example. The direct consumption by man of proteins derived from novel vegetable or animal sources and from micro-organisms or proteins substantially modified by new processes raises considerable concern. Before human studies are undertaken, adequate studies in animals are essential. Some individuals may develop allergic reactions to novel proteins or components. Individuals predisposed to metabolic disorders or hepatic or renal disease may also experience adverse effects following consumption. Hence efficient monitoring for side effects is necessary for a considerable time following exposure of the general population.

3.4 Bio-medical products

There is no agreement on regulation at the Community level of bio-medical products although many developments have been made possible by techniques such as r-DNA and cell fusion. These include interferons, already on the market in Italy and Ireland, vaccines for example, for hepatitis B and influenza, serum albumin, urokinase and many other therapeutic and prophylactic substances.

Biologically active substances are regarded as those products which cannot be completely characterized by chemical and physical tests alone. Bio-medical products must be adequately controlled to ensure quality, safety and efficacy. In this respect, principles successfully applied to previous generations of products have emphasized 'in-process' control, thorough characterization of input material, personnel training and design facilities for purification.

Quality may be assured, as in the United States, by using products that are manufactured according to Current Good Manufacturing Practices (CGMP) and by using CGMP in processing.[42] General guidelines for the control and standardization of biologicals intended for human use have been drafted by the National Institute for Biological Standards and Control (NIBSC) of the United Kingdom, the Food and Drug Administration of the United States and the World Health Organization.

Great flexibility should be exercised in establishing control requirements because these should be influenced by the length of the period over which the product will be consumed, the amount of previous clinical experience, and whether new biotechnology products are identical to previously approved ones. For example, a substance intended for repeated administrations in chronic disease is likely to need rigorous examination for traces of antigenic contaminants. On the other hand different criteria might be required for products designed for the treatment of an acute form of a life-threatening disease.

3.5 Conclusions

Genotypic or phenotypic changes at the various stages of the production process, could cause undesirable properties in the product. The prevention of these changes is in the interest of the manufacturer, wishing to maintain maximum profitability of his process. Close process control and quality control of the product should enable him to recognize hazards related to changes in the culture of the producing micro-organism. This could be assured by using, as in the United States Current Good Manufacturing Practices.

Research is recommended on more convenient and quick methods to detect microbial contamination within large culture volumes and to discover the potential for phenotypic and genotypic changes in micro-organisms.

7 Potential risks from agricultural and environmental applications

The application of biotechnology to increase energy and agricultural production or to improve environmental quality involves the selection or creation of organisms which will have to become established in the environment to carry out their intended function effectively. This has raised concern that the change in the ecosystem directly encountered could be deleterious and/or that the organisms could have negative effects if they become established outside the specific environment for which they were intended.

New organisms can affect the environment in a variety of ways. They can do so directly by replacing or competing with an organism currently in place, by eliminating natural enemies of members of the ecosystem, by altering nutrient flows within the environment, for instance, as a result of effects on soil organisms. The interaction may result from the organism itself or from any of the substances which it produces or the production of which is impeded by the presence of the organism. Harm can result directly through toxicity, predation or pathogenicity. Thus, anticipated risks include such events as the accidental infection of animals by biological pesticides. In addition, indirect effects are possible. Depletion of a nutrient, mobilization of metals or a change in pH, for example, can adversely affect another member of the ecosystem. Past experience with chemical pollutants has indicated that much harm, especially to the environment, may take years to emerge and often cannot be traced to its source. Here, the problem is magnified because organisms reproduce and migrate.

Many risk-scenarios, of both direct and indirect nature, have been foreseen and some are reported in Annex 10. These risks, if any, cannot at present be fully characterized and it is not clear that they are significant either in absolute terms or in relation to the potential benefits. Yet they need to be discussed, put into perspective and regulated on the basis of relevant experience. Conjectural scenarios should not unnecessarily raise fears and lead to over-regulation of this technology.

1. THE ECOLOGICAL APPROACH

From a review of existing literature on the establishment of exotic micro-organisms, plants, insects, birds and mammals and the development of genetic resistance to antibiotics and pesticides (which involves production of new genotypes) the use of an ecological approach is considered to be valid for assessment of environmental risk associated with genetically engineered organisms:

The likelihood of an organism becoming established and persisting in a given environment is an ecological question regardless of the origin and nature of the differences that makes such an organism becoming established exotic or novel. Also, the proportion of the genome which is new is not necessarily correlated with the degree of impact an organism can have on an ecological system. In fact, an introduced organism which is very similar to those with which it must compete may be more likely to survive than an organism which is moderately or very different. In this case 'similar' is meant to imply the range of environmental properties (for example, temperature, moisture level, habitat type, etc.) tolerated by one organism overlaps the range tolerated by another.[43]

The result of introducing alien organisms is determined by both the nature of the invaded environments and the nature of invading organisms. Environments that are particularly vulnerable are those that are isolated, simplified, and/or stressed and disturbed. The limited floras and faunas of remote islands have usually undergone drastic and often permanent modifications when invaded by continental colonizers. Agricultural and other human-created systems provide ready targets for the proliferation of imported insects, pests and weeds, and even mature ecosystems that are subject to perturbation and stress may be unable to check invaders when normal homeostatic mechanisms are disrupted.

Important characteristics for success of an exotic seem to be that it is overly specialized, that it has highly efficient reproduction and dispersal and that it is an aggressive competitor. All of these factors are generally related to an ability to take advantage of whatever resources are available. The success of exotics is furthered by lack of their own predators and parasites, such that the environment they enter has no defenses against their particular characteristics.

The changes caused by an introduced exotic can vary widely. Competition, predation, parasitism or pathogenicity on the part of the invader can result in reduction or exclusion of native forms. These stresses may also contribute to the establishment of other exotics to cause more ecological imbalance. Over the long term, however, a new organism may finally become integrated into a system. This integration is usually marked by a sensible reduction of individuals. The reasons for the decline of exotics, when this occurs, are as obscure as the reasons for their explosive spread in the first place.

1.1 Conclusions

A review of the literature shows that, firstly, the more disturbed, artificial, or simplified an environment is, the less likely it is that a new balance can be established in a reasonable time. Second, even if an exotic becomes integrated and is no longer explosive, the system it entered may have been modified and is perhaps simpler. Sharples, in his report on *Spread of organisms with novel genotypes*, concluded:

> Unfortunately, no quantitative assessment of the risk of ecological damage by genetically manipulated organisms can be made on the basis of existing ecological knowledge. Efforts directed at devising a numerical probability for establishment and disruption based on the number of negative results that have already occurred would not be meaningful or fruitful for a number of reasons. First, what constitutes a negative result is not always clear, and may be a matter of interpretation. Second, very little attention is ever paid to exotic organisms unless they become a problem, but we do not really have any idea how many have been introduced and failed to get a foothold. Third, ecologists do not understand enough about the complex interactions in an ecosystem to be able to predict the outcome of an introduction. There are too many uncontrolled and unknown factors at this point to handle new situations on anything other than a case by case basis.[44]

[handwritten: For less the added complication of un-predictable human behaviour]

2. MICRO-ORGANISMS

Micro-organisms are likely to be of most immediate commercial importance in applied genetics. Several commercial uses of genetically engineering micro-organisms—pollution control, oil recovery, metal leaching and concentration, and pest control—will require deliberate release into the environment. Microbial ecosystems exhibit the same ecological properties and processes as systems of higher organisms. Micro-organisms compete, prey on each other, and modify each other's environment chemically and physically. Stable climax communities of micro-organisms evolve by the same process of selection and succession as do communities of high organisms. Microbial climax communities are usually extremely resistant to invasion by introduced aliens. For example, despite continual addition of diseased tissue, dead bodies, and excrement to soil, or sewage and agricultural debris to water, it is rare for organisms introduced with these media to persist as permanent residents. An established microbial community will usually react in such a way as to reject any aliens or hold them in check.

Indigenous microflora are not, however, always completely stable and environmental perturbation can often allow new species to become dominant. Among stresses shown to bring about ecological upset are nutrients added by man, toxic pollutants introduced into water, and chemical and physical treatments leading to elimination of large numbers of organisms.

The use of antimicrobial agents in medicine, veterinary practice, and agriculture or the release of toxic materials into flowing water constitutes an ecological perturbation that destroys the original homeostatic mechanism and commonly establishes conditions suitable for the dramatic rise in numbers of a species hitherto uncommon in the locality.[45]

2.1 The problem of genetic stability

In bacteria, natural mechanisms exist for transmitting genetic information between species and even genera so that some new traits may be acquired, across taxonomic boundaries, by species which did not possess those traits originally. These unusual properties of mobility are exhibited among other subcellular mechanisms by plasmids, the molecular entities which are the primary vector for transfer of new genes into bacteria in the r-DNA technique. Plasmids are autonomously replicating extranuclear genetic elements. Like the material of the bacterial chromosome, the DNA of plasmids encodes genetic information.

The kinds of information frequently encountered are resistance factors for antibiotic and heavy metals, and genes that specify functions which allow plasmids to transfer from cell to cell. There are also plasmids known to encode production of antibiotics and others which specify properties that cause pathogenicity (virulence plasmids). Plasmids are frequently referred to as dispensable elements, and are thought to govern functions which are non-essential for normal growth and metabolism of the bacterial host. The genes they carry are often for supplementary activities which allow the host's survival in adverse environments.

It is believed[46] that the potential for exchange of plasmids and bacteriophage, virus that infect bacteria, between different species and genera is so large that, at least in bacteria, the species concept may no longer be meaningful. No single organism can evolve in genetic isolation from other organisms with which it shares a common environment. With genetic engineering, an entirely new micro-organism will not generally be produced. Rather, genes for different or enhanced properties will be added to the genome of a natural organism. However, the degree of change in the genome does not necessarily correlate with the magnitude of the impact of such a change.

In addition to the many concerns about these organisms—survival, transport, occupation of ecological niches, genetic stability—that have already been raised about exotic organisms released into new environments, here there is an additional question: whether the newly acquired properties give the engineered organism an undesirable selective advantage over unaltered members of the species.[47]

Despite the remarkable advances in the isolation, analysis, construction and methods of introducing new genes into micro-organisms, there is little information on the potential survival, establishment and growth of *E.coli* K-12 host-vector systems in natural environments. For example, plasmid pBR322 has been used as a vector for transferring engineered genes, as this plasmid is both non-conjugative and poorly mobilizable, hence considered safe (that is, it has a low risk of being transferred to the bacteria indigenous to natural habitats, including that of the human gastrointestinal tract). However, many of the data on gene transfer between bacteria have been obtained with laboratory strains and with genetic recombination studies being performed under optimal laboratory conditions for gene transfer. Such data, however, may not be directly applicable to genetic transfer in situ, that is, in natural aquatic and terrestrial environments. Conjugation between environmental isolates of Gram-negative bacteria appears to occur at temperatures that are suboptimal or even inhibitory for laboratory strains.[48] Information necessary to assess the risk to the biosphere of non-debilitated strains, designed to survive and function for a long time in the environment, appears to be even more scarce.

Direct or indirect changes may affect the environment when deliberate or accidental release of micro-organisms takes place. Many risks have been foreseen but they cannot, at present, be fully characterized because insufficient information is available. The main points of uncertainty are:[49]

(1) the possibility of host-range shifts, for example, of biological pesticides;
(2) the exchange of genetic information and the extent to which this would lead to unanticipated effects;
(3) whether an introduced micro-organism may perturb existing ecological relationships and the impact of such perturbation;
(4) whether the various roles of micro-organisms in the environment are understood.

2.2 Conclusions

There is an acute need for further research to allow adequate assessment of the risks and the formulation of meaningful guidelines. In addition to producing genetically engineered micro-organisms, the biotechnology industry should assess the survival, establishment and growth of, and the genetic transfer by, these engineered organisms under the natural conditions in which they are to be used.

3. PLANTS

Prospects for the generation of improved plants through biotechnology exist in many areas. Specific gene transfer is just one of the available techniques, while less sophisticated ones such as interspecific hybridization, anther and pollen culture, somaclonal variation and protoplast fusion also have important roles to play. The potential for the transfer of specific genes is immense. There are about 250,000 flowering plants (200,000 dicotyledons and 50,000 monocotyledons) each with somewhere about 50,000 genes. These genes are located not only in the nucleus but also in the chloroplasts and mitochondria. Although there will be considerable gene duplication, this means that there in theory will be some 10,000,000 genes available for transfer.

3.1 Research

The new technologies, must be used in combination with classical plant breeding techniques to be effective. Even though species incompatibility can now be overcome, the new plant must still be selected, regenerated from single-cell culture, and evaluated under field conditions to ensure that the new genetic change is stable and the attributes of the new variety meet commercial requirements.

The introduction of new plants into the environment must always be accompanied by careful evaluation and constant oversight by the agricultural research community. This has been dictated in the past by the development of new crop cultivars and the need not only to assess their agricultural stability but also to insure the preservation of the environment. Undesirable properties have usually been detected during experimental trials.

Responsible use of biotechnological products must be insured by the assessment of their biological and environmental impact. Initially risks should be evaluated in the context of individual experiments to utilize available knowledge more effectively. As knowledge and experience in r-DNA advances, broadly applicable protocols and guidelines can be developed.

Field testing of plants appears to be a necessary step in the research process because:[50]

(1) many plants may respond to greenhouse or growth chamber testing differently than they do in field conditions;
(2) it is very difficult in the greenhouse to simulate natural diurnal and stress conditions found in the field;
(3) the number of plants which can be grown in the growth chamber or greenhouse is limited.

A plant obtained through genetic engineering must be observed under its natural growing conditions to determine if the genetic trait which was modified is expressed as expected or predicted and does not have any detrimental or debilitating effects.

The original guidelines of the United States National Institutes of Health (N.I.H.) prohibited 'deliberate release into the environment of any organism containing r-DNA'. The situation evolved with the recognition that genetic engineering systems were being successfully developed to modify plants. Guidelines for field testing of certain modified plants were laid down in the United States in 1983. Approval of experiments was required and conditions under which certain plants could be approved for release were established. Plants covered by the guidelines would be species of a cultivated crop of a genus that has no species known to be a noxious weed. The vector could consist of DNA from

(1) host-vector systems already approved;
(2) plants of the same or closely related species;
(3) non-pathogenic micro-organisms or non-pathogenic lower plants;
(4) plant pathogens if sequences known to produce disease symptoms have been deleted or
(5) a combination of (1) to (4).

The DNA to be introduced would be well characterized and contain no sequences harmful to humans, animals and plants. Also, the investigator was required to develop procedures to assess the spread of plants containing r-DNA.

Agricultural plants have been the most economically important group of exotic introduced organisms in the past. In examining the environmental effects of exotic plants introduced for agricultural purposes, the conclusions have been almost entirely positive. The introduced crops have been minimally disruptive to native plant and animal communities and for the most part are dependent completely on man for their survival. Bates (1956) gives three reasons for this dependence. Some crop species have lost the capacity to produce viable seed, and must be propagated by man vegetatively. Others produce viable seed, but in such a way as to require human dispersal and planting (for example, maize). For most crops, however, it is merely assumed that their failure to become established in the wild is due to their inability to compete. Most introduced plants can only survive in environments in which man-made alterations are maintained by continual disturbance, creating 'open' habitats.[51] Still, the relationships between the major crops and their weedy wild relatives are not well understood. The case of bamboo can be mentioned which apparently spread from cultivation to cover whole mountain sides of Caribbean islands.

Developing new plants would be of little advantage unless they can be integrated profitably into the farming system either by increasing yields and the quality of crops or by keeping costs down.

The three major goals of traditional crop breeding are:[52]

(1) to maintain or increase yields by selecting varieties for
   - pest (disease) resistance,
   - drought resistance,
   - increased response to fertilizers,
   - tolerance to adverse soil conditions.
(2) to increase the value of yields by selecting varieties with such traits as:
   - increased oil content,
   - improved storage qualities,
   - improved milling and baking qualities,
   - increased nutritional value, such as higher level of proteins.
(3) to reduce production costs by selecting varieties that:
   - can be mechanically harvested, reducing labour requirements,
   - require fewer chemical protectants or fertilizers,
   - can be used with a minimum tillage system, conserving fuel or labour by reducing the number of cultivation operations.

3.2 Future developments

Some applications of the new genetic technologies to crop plants, of more relevance to the environment, will be briefly illustrated.

3.2.1 Resistance to environmental stress

The major soil stress faced by plants include insufficient soil nutrients and water or toxic excesses of minerals and salts. The total land area with these conditions approaches 4 billion hectares or about 30 per cent of the land area of the world. Traditionally, through the use of fertilizers, lime, drainage, or fresh water irrigation, environments have been manipulated to suit the plant. Molecular biology might make it easier to modify the plant to suit the environment.

(1) Many micro-organisms and some higher plants can tolerate salt levels equal to or greater than those of sea water. While salt tolerance has been achieved in some varieties of plants, the classical breeding process is arduous and limited. If the genes can be identified, the possibility of actually transferring those for salt tolerance into plants makes the adaptation of plants to high salt, semiarid regions with high mineral toxicities or deficiencies more feasible. In the future, selecting for tissue cultures for metabolic efficiency, for example, cell lines for resistance to salts and for responsiveness to low nutrient levels will become more important.

(2) It is known that in some plants the synthesis of certain proteins is stimulated by drought stress. So far these proteins have not been characterized or their biochemical functions described. It is likely that such proteins are of adaptive significance in helping to conserve water, maintain growth rates during periods of stress and in limiting damage to the metabolic machinery of the plant so that when the stress is reduced normal growth can quickly re-start. The genes for such proteins are possible targets for isolation and transfer to other plants to increase their drought tolerance. In the longer term, there may be some opportunities to improve the water use efficiency of plants, perhaps through manipulating more complex morphological aspects such as root and cuticle structure.

(3) Temperature is another stress syndrome which is related to water stress. Again there is variability in the abilities of plants to grow at high and low temperatures. Work is being done on the isolation of heat and cold-shock proteins. If, as it is likely, they turn out to be of adaptive significance, these small proteins are possible candidates for genetic transfer into crop plants. Less

work is being done on the genetic basis of tolerance to low temperatures but there may also be specific proteins involved for which genetic material could be transferred.

The above applications could be employed in agricultural programmes in less developed countries where, commonly, supplies of fertilizers and lime are scarce, the potential for irrigation is small, and adequate support for technological innovation is limited. In addition, marginal land in Northern Countries could be reclaimed for forest products and biomass.

3.2.2 Macrominerals

Conventional farming uses large amounts of fertilizers, usually a mixture of nitrogen, phosphorous and potassium. The use of such fertilizers has led to the dramatic rise in yields in this century's 'green revolution'.

Not all plants are restricted to obtaining macrominerals such as nitrate and phosphate from the soil. Some plants have developed the ability to fix atmospheric nitrogen through associations with certain micro-organisms. The most important symbiosis in terms of fixing nitrogen is that between legumes and *Rhizobia*. There has been considerable speculation about the possibility of transferring genes for nitrogen fixation (the nif genes) from bacteria to cereals. It is now known that in *Klebsiella pneumoniae* the nif genes are a cluster of seventeen genes. However, at this time the transfer and expression of this number of foreign genes into cereals is a very long distant prospect.

Perhaps a more attractive possibility is the transfer of nodulation (nod) genes from legumes into cereals so that cereals could form a symbiotic relationship with *Rhizobia*. However, not only are the products of the bacterial nif and nod genes required but also, because at least twenty plant genes are involved in nodule development, many more genes would have to be identified. Another problem would be to insure that the genes involved in the regulation of nitrogen fixation were also transferred.

Any reduction in the input of fertilizer into the agricultural system would probably be offset by the reduction in yield through the high energy demands of nitrogen fixation. For the time being there is no immediate prospect of biotechnology being used to make cereals which can fix their own nitrogen. However, the new genetic techniques have been used to design *Rhizobia* with increased nitrogen fixation efficiency. Also, there are very real prospects of improving the fidelity of recognition between the host plant and its corresponding species of *Rhizobium*. This is an important area of research because; at the moment, many associations between leguminous plants and the 'wrong' *Rhizobia* do not lead to active nitrogen fixation.

Another possibility is to improve the performance of nitrate uptake and reduction in cereals. It is estimated that a cereal crop may actually take up less than one third of the nitrogen fertilizer applied to it. The enzyme nitrate reductase is the first committed in the reduction of nitrogen by cereals, it may be possible to improve the activity of this enzyme so that better use is made of applied nitrates.[53]

Another important symbiotic relationship beween plants and micro-organisms is that involving the mycorrhizal fungi. In contrast to the legume/*Rhizobia* symbiosis, mycorrhizal associations are formed by many plants. These fungi are of considerable importance, especially the ectomycorrhizas of forest trees and the vesicular arbuscular of many endomycorrhizas of many crop plants. The mycorrhizal fungi allow more efficient uptake of mineral nutrients from soil. The most important effects appear to be in the uptake of phosphate, but there is evidence that mycorrhizal fungi may be important in the uptake of other ions including trace elements, and that they may help that plant in tolerating certain stresses such as heavy metal ions in the soil, or water deficiency. There may be prospects for the genetic manipulation of mycorrhizal fungi so as to increase the beneficial effect of this symbiosis in the host plant. For example, genes coding for nitrogen fixation or for the production of substances toxic to pests or pathogens could perhaps be inserted. It should also be born in mind that undesirable weeds might benefit from such treatment.

Reducing the amount of chemically fixed nitrogens and the cost of natural gas previously used in the chemical process could be the largest benefit of successful nitrogen fixing by crops. Environmental benefits, from the smaller amount of fertilizer runoff into the water systems, would be increased as well. However, it is difficult to predict when this will become reality, and this for two main reasons: first, knowledge of how plant gene expression is regulated is far from being complete and in the short term only single and small groups of genes will be transferred. Second, genetic techniques are only the tools for change in how society produces its food; financial incentives and pressures and regulations will have major influences on their application.

3.2.3 Herbicide and pest resistance

One of the products likely to enter agriculture in the very short term will be plant varieties with novel herbicide resistance. Tolerance is conferred by single genes which can be easily manipulated. Effective expression of herbicide resistance phenotype in plants in the field may result in reducing the application rates needed to control weeds by allowing optimal timing of application. Also, the decision whether to apply herbicides at all can be made on the basis of a post-emergent survey of the situation. The impact of herbicide-tolerant varieties on the use of agrochemicals will undoubtedly be great. The use of existing herbicides may be broadened, resulting as well in an increase in the product lifetimes of chemicals already in use. Achieving selectivity might increase demand for an effective chemical. Indeed, such seems likely in the atrazine/corn market. Corn is already resistant to atrazine, but residues of the herbicide can damage soya beans planted the following season on land where corn and soya beans are rotated. Using less atrazine on corn reduces damage to soya beans, but the increase in weeds reduces corn yields. If the predominant varieties of soya bean were resistant to atrazine, about two or three times more atrazine would be used on related crop land.⁵⁴ Herbicide resistance in crop plants could therefore lead to a more rational use of herbicides employed in cultivation and result in an increase in the product lifetime. However, this benefit also entails questionable profitability. More likely, herbicide resistance in the short term will be developed to allow more intense use of broad spectrum products.

The herbicide industry expects to become increasingly a commodity rather than a speciality business. The marketing of herbicides coupled with resistant varieties may result in savings to the farmer, resulting from reduced weed control expenses, to be captured by increased margins on sale of seed. This may not, however, be a positive development. A similar compensation mechanism will take place with the development of pest-resistant plants.

Vast amounts of pesticides are now being used in agriculture and new chemicals are continuously being developed as pests become resistant. Biological control agents have, up to now, represented the only valid alternative to the use of pesticides. The best known example of the use of an organism for biological control of insect pests is *Bacillus thuringiensis* (Bt). *Baccilus thuringiensis* is active against lepidopteran larvae in a non-selective way. The EPA has registered it for use on field crops, trees, ornamentals, home vegetable gardens and, more recently, stored products (grain and grain products). Bt subspecies Kurstaki is one of the most important in agriculture and forestry because it kills leaf-eating larval forms of moths (more information in *Science* **219**, no. 4585, 11 February 1983, pp. 715–21).

There are two complementary approaches to the further development and exploitation of Bt through genetic engineering. One is to engineer better strains of Bt which produce a more lethal toxin or a toxin with an altered host range. Another approach is to take the gene which codes for the toxin from the *Bacillus* and to clone this into tobacco as the 'test plant'. New genes arising from the manipulation of new strains of Bt could also be transferred.

There are many other candidates for the isolation and transfer of genes which code for products which kill insects: bacteria, fungi, viruses, other plants or even animals. For example, legume seeds are nutritionally rich and in many cases contain anti-metabolites which reduce insect attack on the seeds. The development of poisons to kill insects is just one of the defence strategies adopted by plants. Another mode is production of chemicals which control insect behavior. For example, the marsh pepper produces a substance, called polygodial, which when sprayed onto crops in small amounts prevents aphids from feeding. Polygodial is produced in plants in only a few enzymatically controlled steps from farnesyl pyrophosphate which is present in plants. If the genes coding for the enzymes involved could be isolated and transferred to crop plants it should be possible to produce plants which are resistant to aphid attack and so are less susceptible to aphid-transmitted viruses.

Genetic engineering can affect not only what crops can be grown but where and how these crops are cultivated. Modified plants to suit difficult environments and faster improvements of under-exploited plants will certainly have important economic results. Public funds should assist the research projects which could produce environmental benefits, because these are of less commercial profitability, and not likely to be undertaken by industry.

3.3 Long-term impacts

3.3.1 *The seed industry*

Seed are the delivery system for improvements in plants made by new genetic technologies. Revenues and profits in the seed business, for this reason, are expected to raise dramatically by the year 2000.

A report on the impact of genetic manipulation on 28 key crops has been produced in the United States after three-year's work by over 25 multinational corporations.[55] Forecasts (see Annex 11) showed that major developments, central to the realization of increased profits include:

- economic production of hybrid wheat and barley seed;
- development of selected vegetables (potatoes, tomatoes, brassica) desired by the consumer or food processor;
- improved agronomic and processing features in sugar crops.

Economies of scale in marketing and the potential advantages conferred by new genetic technologies are leading to consolidation of the seed industry. The focus is shifting from agriculture to pharmaceuticals, energy and processed foods. Shortly after the year 2000, it is expected that no more than a dozen global companies will dominate the entire seed market.

3.3.2 Biological diversity

In the examination of the possible applications to agriculture of genetic techniques, the importance of biological diversity to the future of these potential uses must be stressed. Germplasm is the fundamental resource of biotechnology, and it is conceivable that it could become a limiting resource, preventing full realization of genetic engineering's positive potential. Even if plant breeders adequately understand the amount of germplasm presently needed, it is difficult to predict future needs.

Because pests and pathogens are constantly mutating, there is always the possibility that some plant resistance will be broken down. Even though genetic diversity can reduce the severity of the economic loss, and epidemics might require the introduction of a new resistant variety. In addition, other pressures will determine which crops will be grown for food, fibre, fuel and pharmaceuticals, and how they will be cultivated; genetic diversity will be fundamental to these innovations.

Most of the wild plant varieties come from less than a dozen so-called 'Vavilov centres'. Named after a Russian botanist of the 1920s, the centres are areas that were relatively untouched by ice-age climatic changes and as a result developed a high genetic diversity. They are found mainly in the developing world in the tropics and subtropics. Without wild genes, there would probably not be viable sugarcane or tomato industries anywhere in the world. Production of wheat, maize, barley, potatoes, millet, virtually all major crops, has been increased by selective crossbreeding with wild varieties.

Tropical countries are home to more than 70 per cent of the total flowering plants, but no more than a small fraction has been identified, much less scrutinized for potential contributions to medicine, agriculture and industrial uses. Unfortunately, critical losses have already occurred. It is difficult to examine the amount of permanent cover that is being lost; however, it has been estimated that 40 per cent of 'closed' forest has already disappeared, with 1 to 2 per cent cleared annually. If the highest predicted rate of loss continues, half of the remaining closed forest area will be lost by the year 2000.[56] The significance of this loss is expressed by Norman Myers in his report, *Conversion of Tropical Moist Forests*:

> Extrapolation of figures from well-known groups of organisms suggests that there are usually twice as many species in the tropics as temperatate regions. If two-thirds of the tropical species occur in TMF (tropical moist forest), a reasonable extrapolation from known relationships, then the species of the TMF should amount to some 40 to 50 per cent of the planet's stock of species—or somewhere between 2 million and 5 million species altogether.
> In other words, nearly half of all species on Earth are apparently

contained in a biotope that comprises only 6 per cent of the globe's land surface. Probably no more than 15 per cent and possibly much less—have ever been given a Latin name, and most are totally unknown.[57]

For these lost species, some hope may reside in the unknown. Five years ago, a perennial species of highly disease-resistant wild corn was discovered on a remote Mexican hillside. *Zea diploperennis*, a relative of our sweet corn, has the potential to transform agriculture. Yet a bulldozer could have ended its tenure on Earth. Similarly, a little plant from Madagascar, the rosy periwinkle, was found to contain alkaloids that are now used against leukemia and other cancers.

Because of the demands of genetic engineering, the need to collect, preserve and study diverse populations is now greater than ever before. Efforts to collect germplasm have, in the past, been stimulated by devastating crop epidemics which led to the awareness of crop vulnerability. One of the major causes of crop vulnerability was found to be genetic uniformity of cultivated plant varieties. Of course, a narrow genetic base is necessary to obtain the most uniformly high-yielding seed over a short period of time. But when a single variety dominates the planting of a crop the resulting uniformity causes crop vulnerability to pests. Use of high-yield commercial cultivars results in displacement of indigenous varieties.

Germplasm must be adequately maintained to assure viability, 'working stocks' must be made available to the research and breeding community. The primary objective of storing germplasm is to make the genetic information available to breeders and researchers. However, germplasm collection cannot, itself, maintain biological diversity. Protecting the stability of systems is not synonymous with preserving species diversity because genetic evolution cannot take place in storage shelves. To maintain biological diversity, one has to preserve ecological processes.

Should high-volume biological production of the tunicate antiviral compounds prove desirable, it will only have been possible because those tunicates still existed. Indeed, the ultimate genetic engineering is what occurs daily in nature, conducted by millions of species.

Species may be preserved *in situ* and *ex situ*. Especially in aquatic systems, we depend completely on *ex situ* preservation. The dominant role of the private sector in biotechnology creates the need for the private sector to manage biological diversity. On the one hand, the greater interest of the private sector in germplasm resources will present obstacles to widespread collection, management or sharing of germplasm. On the other hand, the economic dependence of the private sector on germplasm resources may make it possible to utilize private sector capital and technology for more successful management of germplasm.

Germplasm is becoming an issue in many national policies.[58] Iraq refuses to allow wild date palm seed to leave the country. Sudan restricts exports of the tree from which gum arabic is made. Latin American nations have not succeeded in getting strains of wild coffee from Ethiopia. According to a senior official of the United Nations Food and Agricultural Organization, Iran threatens decapitation to anyone illegally exporting wild pistachio nuts. Hopefully, national interests can also be used to facilitate better germplasm management and to increase access to greater resources.

3.3.3 *Crop variability and vulnerability*

Conservation of biological diversity is not the only topic of interest. The new genetic technologies will probably have an impact on crop variability and crop vulnerability. These could be either increased or decreased. In theory, genetic technologies could be useful in developing early warning systems for vulnerability by screening for inherent weaknesses in major crop resistance. The new technologies may also enhance the prospects of using variability and creating new sources of genetic diversity by:[59]

- incorporating new combinations of genetic information during cell fusion;
- changing the ploidy (chromosome) level of plants; and
- introducing foreign (non-plant) material and distantly related plant material.

With the potential benefits, however, come risks. Because genetic changes during the development of new varieties are often used extensively, the new technologies could increase both the degree of genetic uniformity and the rate at which improved varieties displace indigenous crop types. Furthermore, it has not been determined how overcoming natural breeding barriers by cell fusion or r-DNA will affect a crop's susceptibility to pests and disease.

Historically, success and failure in breeding programmes are linked to resistance in pests and pathogens. Plant breeders try to keep one step ahead of mutations or changes in pest and pathogen populations; a plant variety usually lasts only 5 to 15 years on the market. There is some evidence that pathogens are becoming more virulent and aggressive—which could increase the rate of infection, enhancing the potential for epidemics. In fact, there is some evidence that the introduction of novel sources of major gene resistance into commercial cultivars of crop plants has resulted in an increase in frequency of corresponding virulence in their pathogens.[60] This has been reported in Australia with wheat stem rust, barley powdery mildew, tomato leaf mould and lettuce downy mildew. The considerable gene flow in the various pathogen populations can considerably alter the frequency of virulence genes. Furthermore, pressures brought about in the evolutionary process have developed such a high degree of complexity in both resistance and virulence mechanisms that breeding approaches, especially those only using single gene resistance, can easily be overcome.

3.3.4 Conclusions

The impact of genetic technologies on conservation of natural ecosystems, genetic variabilty and crop vulnerability is still an unresolved issue. Answers are by no means complete. Challenges and opportunities abound. Research to help understanding of plant genetics, and continuing international debate on the issues arising are essential.

8 Biotechnology regulation in the Member States of the European Community

Over the past decade, biotechnology has been perceived as a subject of scientific research rather than a commercial activity. In relation to this perception some regulatory schemes have been imposed, characterized by flexibility and relatively minimal constraints, on the basis of the traditional freedom of scientific inquiry. Recommendation 82/472/EEC for registration of work involving r-DNA has been issued within this context.

In the last few years products and processes developed from basic research have approached the point of commercialization. This has been reported by the press and public concern has increased. The first response to this concern has mainly consisted in an effort by the National Governments to appoint the Competent Administrations and to examine the existing regulatory framework applicable to biotechnology products and processes.

The situation in the different member States varies substantially case by case. Biotechnology applications and regulations are developing rapidly and a detailed description of standing and draft regulations would be of limited value. However, some conclusions can be drawn regarding the overall situation of present and future biotechnology regulations.

1. BELGIUM

1.1 Research

The United States National Institutes of Health research guidelines are applied, on a voluntary basis, by universities and the pharmaceutical industry. The EEC recommendation 82/472 on notification of research involving r-DNA has not been adopted officially. However, research projects financed by the 'Fond National pour la Recherche Scientifique' (FNRS) must comply with the EEC recommendation. The decision of the FNRS involves notification to the Health Department of experimental protocols on medical research. No risk evaluation of these protocols is carried out by Government Authorities neither before nor after research work has begun.

1.2 Industrial applications

There are no specific regulations applying to biotechnology processes and products. Existing regulations on medicinal products, including blood serum, vaccines, antigens, hormones, antibodies, disinfectants and pesticides would apply to biotechnology products, that is, require approval for industrial production, registering of products and liability of the industrial pharmacologist.

No specific regulations are in force concerning work with pathogenic organisms. Medical surveillance and blood sampling are compulsory only for workers involved in vaccine production. However, medical surveillance of allergenic problems is supplied by the health departments at the request of individual firms.

1.3 Applications in the environment

According to provisions recently enforced, environmental protection management falls under the competence of the regional administration, but transfrontier pollution problems must still be faced by the central government. Because the risks from environmental release of micro-organisms could cause both local and international problems, there is apparently a need to reach an agreement as to who would be responsible before facing more specific problems. Until now, to the knowledge of government officials, no field tests of genetically engineered organisms have been made.

1.4 Future initiatives

An interdepartmental Ad-Hoc Group was established in March 1985 with the mandate:

(1) to carry out a complete inventory of existing regulations which could relate to biotechnology products and processes, to be completed within one year;
(2) the evaluation of the existing frame of legislation as compared to international provisions or guidelines and the outline of future actions.

2. DENMARK

2.1. Research

According to official responses to a questionnaire by the OECD Ad Hoc Working Group on Safety and Regulation in Biotechnology,[61] the National Research Council in Denmark has set up a Committee on Registration for work with genetic engineering. Guidelines were adopted in which all experiments and applications relating to genetically engineered products are to be registered with the Council. The guidelines are administered on a voluntary basis by the Ministry of Education. According to the same response, Denmark has complied since 30 June, 1984 with the EEC Recommendation 82/472.

2.2 Industrial applications

Although the research guidelines are voluntary for private institutions, it appears that industry has considered them as binding. According to the official responses to the OECD questionnaire, industrial production utilizing r-DNA technology is treated like other biotechnological processes. Under the Environmental Protection Act the Chemical Substances and Products Act voluntary notification to the Ministry of Environment takes place.

Moreover, Denmark, under the Act concerning measures against contagious diseases, has a specific law dealing with the handling of pathogenic organisms. Under this Act, notification and authorization of work with, and transportation of, dangerous pathogens is required. No notification has been received so far for large-scale processes with recombinants that express pathogenicity.

2.3 Applications in the environment

The existing industrial chemical legislation provides for the premarket assessment and general regulation of chemical substances, including living micro-organisms. In addition, Denmark can regulate genetically engineered microbial pesticides under existing agricultural chemical legislation. These authorities provide for the review and registration of all pesticides on a product-specific basis.

Denmark is also expecting to apply the Conservation of Nature Act to environmental applications of biotechnology. Under this act, Denmark can issue regulations concerning structural and functional changes of the natural environment. In the response to the OECD questionnaire, Denmark stated that genetically engineered organisms may be considered, in principle, as species without a natural habitat in their country. In addition, regulations might be issued based on ecological changes which might be anticipated if environmental and agricultural applications of genetically engineered organisms take place. This statute is administered by the Ministry of Environment and is concerned with applications.

The Act concerning administration of hunting of wildlife requires Ministerial approval to release species into places where they do not naturally exist. In the above-cited response, Denmark stated that the release of genetically engineered mammals or birds to the environment could be regulated or banned under this statute. This act is administered by the Ministry of Agriculture and would concern applications.

Other agricultural laws that may apply to genetically engineered products cover the prevention of import and/or spread of infectious diseases among domestic animals, the control of import and spread of plant pests, exotic plant species, noxious weeds and plant pathogens.

2.4 Future initiatives

In October 1983 the Ministry of Interior set up a Committee on the legal aspects of genetic engineering and related biotechnology. In summer 1985 this Committee has finalized a report with several recommendations concerning regulation of genetic engineering, including environmental and agricultural applications. The report proposes three different regulations on:

(1) research work;
(2) industrial processing and products;
(3) products for agriculture.

The Ministry of Environment has financed a project to clarify how to use genetic engineering technology to create a better environment. The results, which will focus on food production, pollution control, alternative raw materials for oil-based chemical production, were expected for June 1985.

3. FEDERAL REPUBLIC OF GERMANY

3.1 Research

In the Federal Republic of Germany guidelines for work with r-DNA are mandatory for institutions financed by Federal and Länder Authorities. Moreover, industry is considering these guidelines as binding.

Notification of intent to work with r-DNA is mandatory for all research work. The OTA report on Commercial Biotechnology said that Germany is probably the country where the most restrictive approval process for public-funded research was applied. Experiments at the two high-risk levels require the entire Commission's approval while those at the second lowest containment level must be approved by one or two individual members of the Commission. Research at the lowest level of containment must not be notified. The Commission for Biological Safety must also authorize the use of host-vector systems not enumerated in the r-DNA research guidelines and may approve reductions in level of containment employed. Sanction for non-compliance with the guidelines is the ability of the government to restrict or withdraw funding for an institution's or a scientist's r-DNA research.

Applications of r-DNA technology in excess of 10 litres are not permitted in Germany, but the Federal Health Office may permit exceptions after special hearings at the Zentrale Kommission für die Biologie Sicherheit (Z.K.B.S.).

3.2 Industrial applications

Germany is currently reviewing the research guidelines and control provisions in order to include protection against risk of large-scale industrial applications.

Specific regulations exist in Germany for notification and authorization for work with dangerous pathogens under the

Federal Communicable Disease Act and the Epizootic Diseases Act. Also, the Plant Protection Act, the Food Law, the Drug Law, legislation regarding job safety and environmental protection would apply to large-scale industrial production.

3.3 Applications in the environment

Germany's guidelines ('Guidelines on the protection against the risks of recombined Nucleic Acids', 4th revised edition of 7 August 1981, Bonn) only cover 'in vitro' recombination of nucleic acids. The guidelines do not permit the release of genetically engineered organisms into the environment unless specifically exempted by the Federal Health Office after review by Z.K.B.S.

In the official response to the OECD questionnaire it was suggested that general environmental media laws, public health-related laws (on a product specific basis), and plant product laws (including control of plant pests, plant pathogens and noxious weeds) will apply to genetically engineered organisms and their products.

3.4 Future initiatives

Germany has established an Interdepartmental Advisory Committee (Z.K.B.S.) charged with developing and implementing guidelines for research in biotechnology. Besides reviewing guidelines for large-scale applications, it will also examine whether and under which conditions a release of genetically engineered organisms may be permitted. Results of the ongoing work should be available by the end of 1985.

4. FRANCE

4.1 Research

In France, no administrative procedures apply specifically to genetically engineered organisms. In the official response to the OECD questionnaire, France stated that it was not a party to any international acts or agreements relating specifically to applications of genetically engineered organisms. However, since 1975, laboratories have been invited to submit their projects on genetic recombination *in vitro* to a Commission of scientists under the authority of the General Delegation for scientific and technical research.

4.2 Industrial applications

In the official response to the OECD questionnaire France cited several laws which could apply to biotechnology: the Act on Classified Installations for Environmental Protection, the Act on the Control of Chemicals, Labour Code, Public Health Code, Act on the Prevention of Fraud in marketing and adulteration of food. The provisions could apply to biotechnology products, technologies or both with the same procedures as for other methods and products. France has no regulation with regard to the handling of dangerous pathogens.

4.3 Applications in the environment

In the response to the OECD questionnaire, France stated that in addition to general chemical oversight authorities, it could regulate genetically engineered microbial pesticides under existing agricultural chemical legislation. This law would be applied on a product-specific basis.

4.4 Future initiatives

France has set up an Interdepartmental Discussion Group on the Safety of Industrial Applications in Biotechnology (A.F.N.O.R.). In the reply to the OECD questionnaire, France said that rigid regulations are unsuitable during the early period of rapidly changing developments in genetic engineering technology. France stated that her existing legislation relating to the biotechnology industry in general is adequate to cover the possibilities of accidental release. Thus, no specific regulation exists and none is foreseen to cover agricultural and environmental applications of genetically engineered organisms.

The Economic and Social Council issued a report in January 1983 on 'perspectives for the Bioindustries', which concluded that risks from biotechnology were of low probability and that more work was needed to define which risks society was willing to face.

The Ministry of the Environment will issue a report on the environmental impact of biotechnology by January 1986. It will review existing regulations and will be the basis for future action.

5. GREECE

In Greece there are no specific regulations related to the research and the use of genetically engineered organisms. The EC Recommendation 82/472 is applied on a voluntary basis and guidelines similar to those of the United States National Institutes of Health are followed in research work.

In 1983 the Government set up an interdepartmental Committee, the 'AD-Hoc Committee on Biohazards' which is responsible for the coordination of biosafety activities. The members of the Committee are delegated by the Ministry for Energy, Natural Resources and Technology, Secretariat of the Committee (the former Ministry for Scientific Research), and by the Ministries of Health, Environment and Commerce. Also, academic representatives are members of the Committee. Public and private-funded research protocols are sent to the Committee. The Committee is considering the need to regulate large-scale industrial applications and deliberate release of genetically engineered organisms, and is following for this purpose the ongoing work of the OECD Ad-Hoc Group on Safety and Regulations in Biotechnology.

6. IRELAND

Research guidelines, similar to those developed by the United States National Institutes of Health, apply to research work on a voluntary basis. An r-DNA Committee has been set up within the National Board for Science and Technology as advisory body for the research work.

In the official response to the OECD questionnaire Ireland said that legislation on pollution control, environmental protection, industrial safety, handling and transportation of chemicals and transport of pathogens (regulation of post and communications) may have implications for genetically engineered organisms used in industry. The Departments of Labour, Health, Environment and Communications would be responsible for the application of these Acts. In the response, no specific authority was cited to regulate the handling of plant pathogens, the use of microbial pesticides or the control of plant pests and noxious weeds.

In the official response to the OECD questionnaire, Ireland said that it has no plans to change its existing statutes in view of the large-scale industrial use or of the applications in the environment of genetically engineered organisms.

7. ITALY

7.1 Research

In 1977 a working group of scientific experts recommended a notification system for r-DNA research *in vitro* and the setting up of an Interdepartmental Committee. Up to now Italy has not adopted regulations or guidelines related to research with genetically engineered organisms. However, in the official response to the OECD questionnaire, Italy stated that changes will take place on the basis of the EEC Recommendation 82/472.

7.2 Industrial applications

In the response to the OECD questionnaire, different laws were cited which could relate to the use of genetically engineered organisms: laws related to limits of exposure to pollutants; safety at work in indoor and outdoor environments; production and marketing of vaccines, sera, toxins; chemical composition, registration and production of drugs; production and marketing of food and beverages, limits for pesticides in foodstuffs. Administrative bodies would be the Ministry of Health, the Istituto Superiore di Sanità and the Local Health Services. The provisions would apply on a product-specific basis.

No specific provisions exist for the handling of human, animal or plant pathogens.

7.3 Applications in the environment

Italy can regulate genetically engineered microbial pesticides under existing agricultural chemical legislation, so that they will be reviewed by the same procedure as developed for conventional pesticide products. In response to the OECD questionnaire, Italy discussed also the use of a variety of laws to regulate genetically engineered products for veterinary use and the measures to control the spread of plant pathogens, plant pests, and noxious weeds.

7.4 Future initiatives

Italy is actually reviewing the current legal system in order to apply the EEC Recommendation 82/472 as a ministerial recommendation. The administrative body which will probably have major responsibility in the field of risk assessment, control and product approval would be the Istituto Superiore di Sanità. The scope of action of this institute covers, among others, medicinals, feed and foodstuffs and environmental hygiene.

8. LUXEMBOURG

In the official response to the OECD questionnaire, Luxembourg indicated that, at present, there were no small- or large-scale activities concerning genetically engineered organisms. In any case, legislation on pharmaceutical specialities and manufactured drugs could apply to medicinals obtained biotechnologically. Also, in describing international acts which have implications for the national regulatory situation, Luxembourg cited the EEC Recommendation 82/472. Luxembourg said it had no plans to change its present legal situation in view of large-scale industrial or agricultural applications of genetically engineered organisms.

9. NETHERLANDS

9.1 Research

Guidelines for research involving r-DNA have been prepared on the basis of the United Kingdom Guidelines and of the National Institutes of Health Guidelines by the Ad-Hoc Committee for r-DNA Activities (Ministry of Housing, Physical Planning and Environment and Ministry of Employment and Social Security). Pursuant to the guidelines, proposals for work with r-DNA must be given approval by the Committee. Observance of the guidelines is on a voluntary basis. In the Netherlands, the Committee makes site inspections which will also apply to large-scale industrial applications.

9.2 Industrial applications

In the response to the OECD questionnaire the Netherlands have indicated that there are guidelines relating to large-scale applications of genetically engineered organisms (June 1985).

In March 1985, the Netherlands decided to regulate new products from genetically engineered organisms under existing legislation for chemical products. Thus r-DNA organisms are mentioned as such in the Chemical Substances Act and the Working Environment Act.

Existing legislation which could relate to industrial biotechnology has been listed in the response to the OECD questionnaire: the Nuisance Act, Safety at Work Act, Working Environment Act, Pesticides Act, Chemical Substances Act, Waste Substances Act, Chemical Waste and Oil Waste Act, Dry Rendering Act, Netherlands Pharmacopea Act, Medicines Act, Veterinary Medicines Act, Plant Diseases Act, Dangerous Substances Act, Commodities Act, Sera and Vaccines Act. The Government decided that the same methods and products under these provisions will be applied to genetically engineered organisms. In general, the procedures consist of prior notification, judgement of the application, conferring or rejecting the necessary licence.

9.3 Applications in the environment

In the response to the OECD questionnaire, the Netherlands said that the guidelines recently published (June 1985) cover research for application in the environment. The Netherlands can regulate genetically engineered microbial pesticides under existing agricultural chemical legislation. In the same response, the Netherlands cited an array of laws related to the regulation of import and dissemination of plant pests, noxious weeds and the control of plant pathogens which could be relevant to the uses of genetically engineered organisms in the environment.

9.4 Future initiatives

In the Netherlands, a technology assessment with emphasis on social aspects of biotechnology is now in progress and a report will be issued in 1986. In the official response to the OECD questionnaire, the Netherlands said that a study of the risks of a deliberate release of r-DNA organisms is under consideration and will include regulation and implementation aspects.

10. UNITED KINGDOM

10.1 Research

A mandatory notification applies to all r-DNA research work. The time of notification differs in relation to the level of risk involved. The United Kingdom has set up a list of low-risk hosts, donors and vectors. Work with low-risk organisms will require notification only one year after the research work has begun. Guidelines on r-DNA research are set by the Genetic Manipulation Health and Safety Commission Advisory Group on Genetic Manipulation from the

Department of Health and Social Security. Government officials, rather than the Advisory Committee, are charged with actual review of project proposals.

In the United Kingdom, risks are determined by considering the survivability and likely harm of the organism containing r-DNA. This method of risk assessment is very different from the systems of other countries, based on the evaluation of the source of the DNA used in the experiment and the pathogenicity of the host-vector system.

Companies in the United Kingdom have to deal with two separate agencies: the Advisory Committee, which promulgates and monitors the rules and the Health and Safety Executive (HSE), which enforces them. The HSE is appointed to inspect facilities for r-DNA research at the two higher containment levels, categories III and IV. For research at these levels the supervising authority must get notice and an opportunity to give advice. Advance notice is required for research at the category II level but not approval. Activities at the category I level can go forward provided only that the local Safety Committee notifies the Central Authorities once a year of new research.

The United Kingdom is the only country in the European Community where guidelines are compulsory for all research and are promulgated under the Health and Safety at Work Act of 1974.

10.2 Industrial activities

Large-scale research in the United Kingdom is treated on a case-by-case basis by the Advisory Committee. Voluntary notification of large-scale work and use in the environment is being required by a separate recommendation. Specific regulations exist regarding notification and authorization for work with, and transport of, dangerous pathogens.

10.3 Applications in the environment

In the response to the OECD questionnaire, the United Kingdom cited the Advisory Committee's Note 13 which states that all experiments involving the genetic manipulation of plant pests (and the use of such genetically manipulated plant pests) requires a licence from the Ministry of Agriculture, Fisheries and Food. Also, the United Kingdom said that the Advisory Committee intends to produce a comprehensive guidance note on 'deliberate release' by the end of 1985. Moreover, the Pesticides and Safety Precautions Scheme (PSPS) is in the process of being replaced by enabling legislation. If enacted, this new pesticides legislation would make compulsory certain PSPS Provisions that may relate to genetically engineered micro-organisms.

10.4 Future initiatives

The United Kingdom Ministry of Environment is planning to organize a meeting on the hazards involved with the use of genetically engineered organisms in the open environment by the end of 1985. Also, the United Kingdom authorities are considering the opportunity of requiring mandatory notification for large-scale industrial applications and for experiments with genetically engineered organisms in the open environment. At present, case-by-case review is seen as the only system which can guarantee safety. General guidelines cannot be drawn up in the near future because experience is lacking.

11. CONCLUSIONS

Member Countries of the European Community have differing regulatory situations in accordance with different levels of experience in the new techniques. This means that level of expertise on evaluation procedures and the attitudes established in respect to environmental risks are equally different.

In the United Kingdom and in Denmark regulations have been set up for observance of notification and safety guidelines in research work with r-DNA.

In the Federal Republic of Germany notification is compulsory for research work supported by the Government while it is voluntary for research funded privately.

In the Netherlands, France, Greece and Ireland notification applies on a voluntary basis and review of protocols take place.

In Belgium, notification is voluntary for medical research funded by the Government but review of protocols is not conducted.

In Italy and Luxembourg no specific recommendation applies to biotechnology research or its developments.

From the viewpoint of promoting international trade and a common market within the Community, harmonization of biotechnology regulations is necessary, to prevent any temptation for countries to 'under-cut' one another with less stringent regulations. At present, no specific provisions exist for the use in the environment of genetically engineered organisms.

9 Biotechnology regulation in the Community

1. BACKGROUND

In the Communication to the Council COM(83)672 of 3 October 1983 the Commission recognized the need for concertation of the activities relevant to biotechnology:

The development of biotechnology in Europe depends upon the separate decisions of many independent actors—scientific researchers, financial investors, corporate management, final consumers—whose individual decisions can be indirectly, but often decisively, influenced by the actions of the public authorities. The creation of a context favourable and encouraging for the development of biotechnology in Europe demands some coherence in these actions. This presents the challenge not only of coherence between the Community and the Member States, but also between separate Ministries and Agencies within each State, and across the services of the Community institutions. To this challenge, industry has responded by forming alliances, consortial activities and by participating actively in the working group of the European Federation of Biotechnology. At national level, many countries have created interdepartmental groups or special 'missions' to integrate and mobilize national efforts.

In the same Communication six priority areas were outlined for action to stimulate biotechnology in the Community and to increase the competitiveness of Europe's bio-industries:

(1) research and training;

(2) monitoring and concertation of biotechnology policies;
(3) new regimes on agricultural outputs for industrial use;
(4) a European approach to regulations affecting biotechnology;
(5) a European approach to intellectual property rights in biotechnology;
(6) demonstration projects.

With respect to actions 1 and 2 the Council adopted a multiannual research programme (COM(84)230 final) with a Decision of 12 March 1985. Also, a 'Management and Coordination Advisory Committee' composed of public servants from the Member States was established to oversee the implementation of the Biotechnology Action Programme.

2. THE NEED FOR HARMONIZED REGULATIONS IN THE COMMUNITY

The Commission's FAST programme had emphasized in 1982 that one of the three policy areas of special importance was 'the creation of a common regulatory environment (and hence more truly a common market) within the Community'.

Industrialists—particularly in the pharmaceutical industry—expressed concern about the prospects of excessive restriction limiting their scope in research and hence their competitiveness: Sweden, the Netherlands and Japan were countries felt by some of their industrialists to have adopted an excessively conservative position on r-DNA work, to the detriment of their biotechnology, and to have been slow in following the consensus for relaxation.

In the Safety in Biotechnology Working Group of the European Federation of Biotechnology, industrialists and academic researchers have joined to draft 'recommended guiding principles'.

The OECD, through its Committee on Science and Technology Policy, has instituted a major study of government policy on the control of biotechnology in its member states. The AD-Hoc Group of the OECD, set up in December 1982, has the priority task, in view of international harmonization, to identify a small number of groups of DNA bases, particularly those coming from plasmids already widely used in research, which would not cause additional risk if they were produced on a large scale by fermentation. This work could lead to the production of a list of groups of DNA bases for which recombination could be authorized in industrial production.

In the European Parliament numerous questions have been asked and a Biotechnology Hearing was held in November 1985 on the Impact of Biotechnology and its international political implications.

The Parliamentary Assembly of the Council of Europe adopted on January 1982 recommendations on genetic engineering (Recommendation 934 (1982)). Following this Recommendation, the Ministers of the Council of Europe adopted a Recommendation on notification of research work with recombinant DNA (Recommendation no. R(84)16).

The Council of Ministers adopted in June 1982 a Recommendation (84/472/EEC of 30 June 1982) on the problems arising in safety related to r-DNA work. It envisages:

- the registration by the competent authority in each Member State of every laboratory using recombinant DNA;
- the notification of protective measures and controls;
- a description of the programme of research which enables the risk to be evaluated.

On 2 February 1984 a decision of the Commission established the Biotechnology Steering Committee (BSC), to coordinate the implementation of the Commission's initiatives in the biotechnology, as described in the October

1983 Communication to the Council, 'Biotechnology in the Community'. It comprises the Directors-General of the services principally concerned with biotechnology, which during 1984/85 have included the following Directorates-General:

  III—Internal Market and Industrial Affairs;
  IV—Competition;
  VI—Agriculture;
  XI—Environment, Consumer Protection and Nuclear Safety;
  XII—Science, Research and Development;
  XIII—Information, Market and Innovation.

The Work of the Steering Committee is supported by that of several interservice working groups and of the Concertation Unit for Biotechnology in Europe (CUBE).

On 24 July 1985 the BSC established the Biotechnology Regulations Inter-service Committee (BRIC) whose tasks will be:

(1) to review the regulations applied to commercial applications of biotechnology;
(2) to identify existing laws and regulations that may govern commercial applications of biotechnology;
(3) to review the guidelines for r-DNA research;
(4) to clarify the regulatory path that products must follow;
(5) to determine whether current regulations adequately deal with the risks that may be introduced by biotechnology and to initiate specific actions where additional regulatory measures are deemed to be necessary;
(6) to ensure the coherence of the scientific data which will form the basis of risk assessment and in particular to avoid unnecessary duplication of testing between various sectors.

In 1980 The Commission's FAST programme launched a 'Social dimension of Biotechnology' multidisciplinary working group (1980–2) which offered five general principles to guide public (and Community) policy in the management of the 'bio-society':

(1) there can be no policy of zero risk;
(2) there are limits to the competence of experts;
(3) even within these limits, their credibility/acceptability is questioned;
(4) there is a need for adaptive strategic management, taking account of long-term, 'total system' effects so far as currently perceivable, and accepting that the perceptions will be modified by experience;
(5) there is a need for continuing education at many levels, including decision makers, the general public and the scientists themselves.

In fact, these 'general principles' highlight one of the major problems which is encountered in approaching biotechnology regulation: the *scientific uncertainty*.

Paradoxically, regulatory decisions in risk prevention are likely to be most urgent where scientific knowledge is the most uncertain. The origins of regulatory problems guarantee their relative insolubility because they are typically discovered as unwanted side effects. Those who operate the process would normally not be expected to be competent in sciences needed to understand the health and environmental hazards. The scientific bases needed tend to be fragmentary or non-existent, drawing on relatively underdeveloped fields such as ecology or toxicology. The data that do exist have often been produced for other purposes and are therefore less appropriate and reliable when used in a different context. Even methods of investigation may be uncertain and open to debate.

Scientists normally prefer a more manageable amount of uncertainty, while politicians and legal experts want none; regulation, however, always has to deal with uncertainty.[62] There is a challenge inherent in attempting to regulate biotechnology: to focus research on key regulatory issues like intrinsic hazards, survivability and monitoring to provide guidance to regulators. In the meantime, uncertainty should not become a justification for regulatory paralysis.

10 The existing regulatory framework for protection of the environment

The first issue to be considered is the extent to which the existing regulatory framework applies or can be adapted to assess risks and benefits associated with biotechnology. The benefits of the technology should be realized while minimizing the risks of an adverse environmental impact.

To simplify the discussion, the present legislation will be examined in the light of a limited number of issues which are of particular concern to the potential environmental risks from biotechnology. These include:

- the assessment of environmental impact and risks from the use in the open environment of biotechnology products;
- the assessment of environmental impact and risks from accidental or inadvertent release of living organisms.

1. ASSESSMENT OF THE ENVIRONMENTAL IMPACT AND RISKS FROM THE USE IN THE OPEN ENVIRONMENT OF PRODUCTS DERIVED FROM BIOTECHNOLOGY

It is generally accepted that, prior to the introduction into the environment of potentially harmful substances, there is need for some degree of assessment of their impact.

Broadly speaking, products of biotechnology can be divided into:

- living organisms

- inanimate substances which can be produced by living organisms.

The main difference between these two categories is, of course, *the ability to reproduce*.

1.1 Council Directive 79/831/EEC

In 1967 the Council of the European Communities adopted a first Directive in order to approximate laws, regulations and administrative provisions in the Member States relating to the classification, packaging and labelling of dangerous substances. This Directive has been amended six times over subsequent years. The 6th Amendment (Council Directive 79/831/EEC of 18 September 1979) is the version that is in force today.

The 6th Amendment contains provisions relating to the pre-marketing notification of new substances and to the classification and labelling of both new and existing substances. Existing substances are those which were on the Community market before 18 September 1981. These substances will be listed in the EINECS inventory, which is currently being prepared.

All other substances are new substances and must be notified before they are placed on the market. The notification dossier must contain information necessary to evaluate the risks which the new substances may present for man and the environment and a proposal for classification and labelling of the new substance.

This classification and labelling proposal is confirmed or amended by the so-called 'Technical Progress Committee' (TPC). This is a Committee of representatives of the Member States whose purpose is to adapt the 6th Amendment to technical progress. All new dangerous substances are added to Annex 1 of the 6th Amendment after the vote of the TPC. Article 2 defines substances as:

... chemical elements and their compounds as they occur in

natural state or as produced by industry, including any additives required for the purpose of placing them on the market.

This definition clearly applies to all inanimate substances which can be produced biotechnologically. Whether living organisms can be defined as 'a compound of chemical elements' remains, at this stage, only a political problem. In fact this definition appears, for this purpose, *to be incomplete but not incorrect*. Supposing living organisms were susceptible to regulation under the 6th Amendment we will have 'new substances' (that is, not in the Community market before 18 September 1981) notified under Article 6 of the Directive. Article 6 requires the manufacturer or importer of a new substance to submit a notification 45 days before the substance is placed on the market. The notification must contain:

- a technical dossier supplying the information necessary for evaluating the foreseeable risks, whether immediate or delayed, which the substance may entail for man and the environment, and containing at least the information and results of studies referred to in Annex VII, together with a detailed and full description of the studies conducted and of the methods used or a bibliographical reference to them;
- a declaration concerning the unfavourable effects of the substance in terms of the various uses envisaged;
- the proposed classification and labelling of the substance in accordance with this directive;
- proposals for any recommended precautions relating to the safe use of the substance.

The kind of information required in Annex VII to identify the substance (and the foreseeable risks), however, does not appear relevant to living organisms.

ANNEX VII

INFORMATION REQUIRED FOR THE TECHNICAL DOSSIER ('BASE SET') REFERRED TO IN ARTICLE 6 (1)

When giving notification the manufacturer or any other person placing a substance on the market shall provide the information set out below.

If it is not technically possible or if it does not appear necessary to give information, the reasons shall be stated.

Tests must be conducted according to methods recognized and recommended by the competent international bodies where such recommendations exist.

The bodies carrying out the tests shall comply with the principles of good current laboratory practice.

When complete studies and the results obtained are submitted, it shall be stated that the tests were conducted using the substance to be marketed. The composition of the sample shall be indicated.

In addition, the description of the methods used or the reference to standardized or internationally recognized methods shall also be mentioned in the technical dossier, together with the name of the body or bodies responsible for carrying out the studies.

1. IDENTITY OF THE SUBSTANCE

1.1. Name

1.1.1. Names in the IUPAC nomenclature

1.1.2. Other names (usual name, trade name, abbreviation)

1.1.3. CAS number (if available)

1.2. Empirical and structural formula

1.3. Composition of the substance

1.3.1. Degree of purity (%)

1.3.2. Nature of impurities, including isomers and by-products.

1.3.3. Percentage of (significant) main impurities

1.3.4. If the substance contains a stabilizing agent or an

inhibitor or other additives, specify: nature, order of magnitude: .... ppm; ....%

1.3.5. Spectral data (UV, IR, NMR)

1.4. Methods of detection and determination

A full description of the methods used or the appropriate bibliographical references

3. PHYSICO-CHEMICAL PROPERTIES OF THE SUBSTANCE

3.1. Melting point
.... °C

3.2. Boiling point
.... °C .... Pa

3.3. Relative density
.... (D_4^{20})

3.4. Vapour pressure
.... Pa at .... °C

.... Pa at .... °C

3.5. Surface tension
.... M/m (.............. °C)

3.6. Water solubility
.... mg/litre (.............. °C)

3.7. Fat solubility
Solvent—oil (to be specified)

.... mg/100 g solvent (.............. °C)

3.8. Partition coefficient

n-octanol/water

3.9. Flash point

.... °C ☐ open cup ☐ closed cup

110 Protection of the environment

3.10. Flammability (within the meaning of the definition given in Article 2 (2) (c), (d) and (e))

3.11. Explosive properties (within the meaning of the definition given in Article 2 (2) (a))

3.12. Auto-flammability

.... °C

3.13. Oxidizing properties (within the meaning of the definition given in Article 2 (2) (b))

4. TOXICOLOGICAL STUDIES

4.1. Acute toxicity

4.1.1. Administered orally

LD_{50}................ mg/kg

Effects observed, including in the organs ....

4.1.2. Administered by inhalation

LC_{50} .... (ppm) Duration of exposure .... hours

Effects observed, including in the organs ....

4.1.3. Administered cutaneously (percutaneous absorption)

LD_{50}................ mg/kg

Effects observed, including in the organs ....

4.1.4. Substances other than gases shall be administered via two routes at least, one of which should be the oral route. The other route will depend on the intended use and on the physical properties of the substance.

Gases and volatile liquids should be administered by inhalation (a minimum period of administration of four hours).

In all cases, observation of the animals should be carried out for at least 14 days.

Unless there are contra-indications, the rat is the preferred species for oral and inhalation experiments.

The experiments in 4.1.1, 4.1.2 and 4.1.3 shall be carried out on both male and female subjects.

4.1.5. Skin irritation

The substance should be applied to the shaved skin of an animal, preferable an albino rabbit.

Duration of exposure .... hours

4.1.6. Eye irritation

The rabbit is the preferred animal.

Duration of exposure .... hours

4.1.7. Skin sensitization
To be determined by a recognized method using a guinea-pig

4.2. Sub-acute toxicity

4.2.1. Sub-acute toxicity (28 days)

Effects observed on the animal and organs according to the concentrations used, including clinical and laboratory investigations ....

Dose for which no toxic effect is observed ....

4.2.2. A period of daily administration (five to seven days per week) for at least four weeks should be chosen. The route of administration should be the most appropriate having regard to the intended use, the acute toxicity and the physical and chemical properties of the substance

Unless there are contra-indications, the rat is the preferred species for oral and inhalation experiments.

4.3. Other effects

4.3.1. Mutagenicity (including carcinogenic pre-screening test)

4.3.2. The substance should be examined during a series of two tests, one of which should be bacteriological, with and without metabolic activation, and one non-bacteriological.

Under Article 19 Annex VII is not included among the Annexes which can be adapted to technical progress by the TPC. Practically, this means that to modify Annex VII the Council would have to adopt a seventh amendment to Directive 67/546/EEC.

Finally, Article 8 exempts from notification 'substances placed on the market for research or analysis purposes in quantities of less than one tonne per year per manufacturer or importer and intended solely for laboratories'. This requirement does not appear to offer enough guarantees in respect of the prevention of risks to human health and the environment. It is sufficient to consider the case of self-replicating micro-organisms able to cause harm and not contained biologically nor physically. In this case there is no quantity small enough to be considered safe.

1.1.1 Conclusions

The prevention and control of hazards when biotechnology products consist in living organisms is unlikely to be effective using Directive 79/831/EEC. It is thought that the ambit of application of Directive 79/831/EEC to biotechnology is limited to all inanimate substances produced by living organisms and not falling under the categories of products exempted and listed in Article 1.

1.2 Council Directive 76/768/EEC

The Directive on the approximation of the laws of the member States relating to cosmetic products has the objective to protect human health when cosmetics are applied under normal conditions of use.

Under the Directive,

- products to be considered as cosmetic products are listed in Annex I;
- prohibited substances are listed in Annex II;
- a positive list of substances which can be used within

certain limits and in certain conditions is set up in the first part of Annex III;
- a positive list of colouring agents is set up in the second part of Annex III;
- a list of substances which can be used within certain limits and conditions (as in Annex III); but which are subject to review by a Committee for Adaptation to Technical Progress is set up in Annex IV (these substances should, with time, either enter Annex II or Annex III);
- a positive list of conservants agents is set up in Annex VI;
- a positive list of solar filters is set up in Annex VII;
- two more positive lists are envisaged which will concern anti-oxidants and hair dyes.

Moreover, the Directive sets provisions for product labelling, packaging and advertising and sets a Committee for the Adaptation to Technical Progress of the Directive. This Committee also determines testing methods and the criteria of microbiological and chemical purity for cosmetic products.

1.2.1 Conclusions

Directive 76/768/EEC covers certain cosmetic products obtained by biotechnology. Whenever a substance produced biotechnologically will enter a cosmetic product (marketed in more than one member State of the European Community) as a colourant, conservant, solar filter, anti-oxidant and hair dye not yet listed in Annex, it will have to be approved by the Committee for the Adaptation to Technical Progress and inserted in the corresponding lists. If the substance will not have one of these functions it will not have to be approved.

2. THE ASSESSMENT OF ENVIRONMENTAL IMPACT AND RISKS FROM ACCIDENTAL RELEASE OF LIVING ORGANISMS

In addition to the need to control hazards resulting from the planned release of biotechnology products, there is the question of dealing with accidental and inadvertent release of new life forms. Such releases would include, among others, major spills from otherwise 'contained' applications of biotechnology, and emissions under normal conditions of plant operation such as aerosol or viable wastes discharged in water streams.

The concern here is that the unknown or unexpected injury to people and the environment could occur before the organisms could be contained and cleaned up.

2.1 Council Directive 82/501/EEC

The Directive on the major-accident hazards of certain industrial activities was designed for two purposes:

(1) to reduce, as early as the design stage, and throughout the operation of the plant facilities, the probability of such accidents occurring, by studying possible causes, monitoring critical points, anticipating the combinations of events which might lead to an accident, and introducing more stringent safety measures.
(2) to prevent such accident—should one occur—from turning into a disaster by limiting the consequences as much as possible. Control and safety machinery must be set up and emergency plans prepared.

The Directive defines a major accident as a major emission, fire or explosion resulting from uncontrolled developments in the course of an industrial activity, leading to a serious danger to people and/or the environment and involving one or more dangerous substances.

The risk of a major accident depends essentially on four factors:

(1) the nature of the substance in question;
(2) their quantity;
(3) the technological process;
(4) the location of the industrial activity.

The Directive covers industrial activities where two conditions are met:

(1) the presence—ascertainable or recognized as possible—of dangerous substances in the industrial activities as products necessary for the technological process, manufacturing product, by-products or residues.
(2) the technological processing carried out which must be listed in Annex I of the Directive, and includes all basic industrial processes of the chemical industry.

The Directive defines a substance as dangerous when it falls into one of the four categories listed in Annex IV of the Directive: very toxic, toxic, flammable and explosive. The mechanism of the Directive can be divided into two parts which correspond to two different fields of application and sets of provisions.

The first part sets up a regulatory framework in which any person in charge of an industrial plant (where the two above cited conditions apply) is required to prove to the competent authorities, at any time, that they have identified major-accident hazards, adopted the appropriate safety measures, and provided the people working on the site with information, training and equipment in order to ensure their safety.

The second part identifies, on the basis of a list of chemicals in Annex III certain industrial activities that are or may be present in excess of a fixed quantity. The systematic control under the second part of the Directive is based on a notification procedure whereby the manufacturer submits to the national competent authorities a detailed study, containing information:

- about the substances and processes involved, forms in

which they may occur in case of foreseeable irregularities, risks entailed and emergency methods and precautions;
- about the installations, siting, exposed groups, sources of hazard and preventive measures;
- about possible major hazard situations, emergency plans, alarm systems and resources for dealing with major accidents.

The application of this Directive to industrial biotechnology meets several limiting factors:

(1) In Annex I are listed the chemical processes to which the Directive applies. These processes can also be carried out by micro-organisms. However, if Annex I mentioned in a more explicit way 'the processing of substances using biological agents' the applicability of this Directive to biotechnology would be more clear.
(2) The definition attributed to dangerous substances, substances which are toxic, very toxic, flammable or explosive, clearly covers only some of the substances which could be involved in industrial biotechnology.

Supposing that these limits do not exist, and only making reference to the 'chemicals which can be produced using r-DNA technologies' (Annex 6 of this report) we would find acetone (highly flammable), ethanol (highly flammable), ethylene oxide (extremely flammable), methanol (highly flammable) and propylene oxide (extremely flammable). Among these, ethylene oxide and propylene oxide would require notification if present in quantities equal or above 50 tonnes; acetone, ethanol and methanol if present in quantities above 50,000 tonnes.

For quantities of dangerous substances below those indicated in Annex III the only provision applicable would be that indicated in the first part of the Directive. Moreover, if substances not classified as dangerous were produced with the large-scale use of micro-organisms dangerous to people and the environment but with no acute toxicity under Annex IV the industrialist would not even be requested to prove that major-accident hazards had been identified.

If Directive 82/501/EEC was to cover major-accident hazards from industrial biotechnology the definition of dangerous substances should cover biological agents dangerous to the environment; therefore the indicative criteria of Annex IV should include a new criteria, for example, substances dangerous for the environment where the possibility exists of causing transmittable disease in people, animals and plants. Moreover, Annex II (stored substances to be notified) and Annex III, the list of substances where the notification applies, should also be amended: it should include, for example, micro-organisms having been identified to cause transmittable disease in people, animals and plants. Also, and above all, Annex I should be amended to include 'bio-processes'.

However, even in this case, there will still be some difficulties with regard to efficient prevention of hazards: in the case of new micro-organisms not known to cause harm there will be no mandatory provisions for the industrialist to identify major-accident hazards, to adopt appropriate safety measures and provide information, training and safety equipment to workers.

2.1.1 Conclusions

Directive 82/501/EEC as such is incomplete with respect to the assessment of the risks of major accidents in biotechnology industry. For a more adequate prevention Annex I, Annex II, Annex III, Annex IV and Annex V would need to be amended. However, while Annex V can be modified by the Committee responsible for adaptation to technical progress set up in Article 15; Annex I, Annex II, Annex III and Annex IV can be only modified by decision of the Council of Ministers.

2.2 Council Directive 85/337/EEC

The Directive for the assessment of the environmental effects of certain public and private projects was formally adopted by the Council on 27 June 1985.

The main elements of the Directive are as follows. A developer, whether in the public or in the private sector, of a project which is likely to have a significant effect on the environment, is required to submit information on the project and its environmental effects to the competent authority, which makes the information available to other public authorities with environmental responsibilities and to the public concerned. The competent authority is obliged to take into consideration the information and opinions it receives, when making its decision concerning the authorization of the proposed project.

The Directive applies to projects which are likely to have significant effects on the environment by virtue, inter alia, of their nature, size or location. Such projects are identified in two lists, which appear as Annex I and Annex II.

Annex I contains projects which are always likely to have significant effects on the environment and which should be assessed in all cases. These projects include crude oil refineries, thermal power stations, installations for storage and disposal of radioactive wastes, works for melting of cast-iron and steel, extraction of asbestos, integrated chemical installations, motorways, trading ports and installations for treatment of toxic and dangerous wastes.

Annex II contains a much longer list of projects in agriculture, extractive industry, manufacturing, energy production, infrastructure and waste disposal. It includes also modifications to Annex I projects. These are projects which may or may not have significant effects, depending on the circumstances. The Directive states that these projects shall be assessed 'where Member States consider that their characteristics so require'.

The first step in the assessment process is the provision of information, which is the responsibility of the developer, which is specified in Annex III: it includes a description of the project and of the environment to be affected, alternatives studied by the developer, significant effects, mitigation measures and a non-technical summary. Some of this information must be furnished where appropriate and some must always be provided.

The second principal step is the process of consultation, both of other authorities and of the public concerned. The information submitted by the developer is made available to authorities with environmental responsibilities and to the public concerned, to give them the opportunity to express their opinion before the project is initiated.

Finally, all the information and opinions received must be taken into consideration by the competent authority in making its decision concerning the authorization of the proposed project. Information and opinions on other matters are taken into account also: the result of the environmental impacts assessment (E.I.A.) is not necessarily the only, or the determinating factor in reaching a decision. The environmental impacts assessment provides an orderly process for gathering and evaluating information and opinion about the likely environmental consequences of proposed projects, to assist decision making.

Annex II of the Directive (projects requiring E.I.A. if the member state so requires) includes certain projects where biotechnology can be used in the industrial processing: extractive industry, processing of metals, chemical industry, food industry and waste disposal. Also, the application of biotechnology in many of the above sectors can involve the release of new and exotic micro-organisms into the open environment (waste processing, metal extraction or aerosols in air emissions from 'contained uses'). In these cases, the information required in Annex 3 should always be required and should also make reference to the specific organism used in the process.

In other words, an environmental impacts assessment

should be mandatory for the developer of projects involving large-scale industrial use or use in the open environment of new or exotic micro-organisms.

2.2.1 Conclusions

Directive 85/337/EEC covers some applications of biotechnology but is not stringent enough in respect of the foreseeable risks of large-scale applications and applications in the open environment of biotechnology.

2.3 Council Directive 75/442/EEC

The Council Directive of 15 July 1975 on waste is a frame Directive which sets out several basic principles of waste management with the general aim of protecting human health and the environment:

(1) when possible, wastes should be recycled in order to conserve natural resources;
(2) waste must be disposed of without risk to water, air, soil, plants and animals, and without causing a nuisance through noise and odours;
(3) a system of permits should be set up by local authorities for undertakings which treat, store, tip or collect wastes;
(4) local authorities should draw up plans relating to waste management and disposal;
(5) the 'polluter pays' principle must apply in waste disposal.

The Directive defines wastes as any substance or object which the holder disposes of or is required to dispose of pursuant to the provisions of national laws in force.

The scope of the Directive excludes:

- radioactive wastes;
- mining wastes;
- animal carcases and agricultural wastes of faecal origin;
- waste waters with the exception of waste in liquid form;

- gaseous effluents emitted into the atmosphere;
- waste covered by specific Community rules.

The scope of this Directive clearly covers wastes which could arise from industrial applications of biotechnology.

2.4 Council Directive 78/319/EEC

The Council Directive of 20 March 1978 on toxic and dangerous waste, within the frame of Directive 75/442/EEC on waste, sets provisions for a system to monitor all installations which produce, hold or dispose of toxic and dangerous waste, to keep records regarding disposal, to ensure that the carriage of toxic and dangerous waste is accompanied by an identification form and that programmes are drawn up in which the various waste disposal operations are taken into account.

A Committee on Adaptation to technical progress is established to update the list of the toxic and dangerous wastes in the Annex to the Directive. In adding a substance of unknown danger at the time of the notification of the Directive, the Committee shall take into account (Art. 17) 'the immediate or long-term hazard to man and the environment' presented by the waste by reason of its toxicity, persistence, bioaccumulative characteristics, physical and chemical structure and/or quantity. The directive defines toxic and dangerous waste as 'any waste containing or contaminated by the substances or materials listed in the Annex' in a nature, quantity or concentration to constitute a risk to health or the environment.

The Directive does not apply to:

- hospital wastes;
- effluents discharged into sewers and water courses;
- emissions to the atmosphere;
- household wastes;
- mining wastes;
- radioactive wastes;

- animal carcases and agricultural waste;
- animal carcases and agricultural wastes of faecal origin;
- explosives;
- other toxic and dangerous wastes covered by specific Community rules.

At present the Directive can cover wastes from certain biotechnology research activities because in point 20 of the Annex are included chemical laboratory materials, not identifiable and/or new, whose effects on the environment are not known. Pharmaceutical compounds too are included in the Annex.

Biologically active wastes are not excluded from the scope of this Directive and in principle those containing high concentrations of dangerous pathogens could be added to the Annex by the Committee of Adaptation to technical progress. However, the Directive does not appear relevant to other biologically active wastes because these wastes are in liquid form and will generally be discharged into sewers (exempted from the scope of this Directive).

2.5 Council Directive 84/631/EEC

Council Directive of 6 December 1984 on the transfrontier shipment of hazardous wastes applies to the same waste defined by the Directive 78/319/EEC on toxic and dangerous wastes. Thus, the same considerations as above are applicable.

2.6 Council Directive 76/464/EEC

The Council Directive of 4 May 1976 is designed to control pollution from the discharge of dangerous substances into the aquatic environment of the Community. The Directive establishes two lists of dangerous substances, for which limited values of emission standards are set by the Council.

List I contains dangerous substances for which authorization is required prior to discharge and for which contamination must be eliminated. List II contains substances which have a deleterious effect on the aquatic environment but which can be confined to a given area related to the characteristics and location of the water into which they are discharged. Pollution from these substances must be reduced.

List I and List II do not make reference to biologically active substances. The limit values of the two lists are based on criteria of toxicity, persistency and bioaccumulation. Moreover, these limit values shall be determined by:

- the maximum concentration of a substance permissible in a discharge;
- where appropriate, the maximum quantity of such a substance expressed as a unit of weight of the pollutant per unit of the characteristic element of the polluting activity;
- where appropriate, limit values applicable to industrial effluents shall be established according to sector and type of product.

This Directive, at present, does not apply to effluents from biotechnology installations. Moreover, its structure does not appear adequate for such objectives for several reasons:

- criteria to establish the limit values for emissions are not applicable to living organisms;
- the approach of the Directive is to cure pollution and not to prevent it;
- no Committee for adaptation to technical progress is established by the Directive and all amendments have to involve a unanimous decision of the Council.

2.7 Council Directive 84/360/EEC

Council Directive of 28 June 1984 on air pollution from industrial plants introduces a system of prior authorizations for the operations and substantial alteration of stationary industrial plants which can cause air pollution.

- Annex I sets a list of plants likely to cause major air pollution;
- Annex II sets a list of most important polluting substances.

The Council will fix emission limit values and the measurement techniques and methods. Directive 84/360/EEC as such does not cover risks which can arise from aerosol emissions of non-debilitated micro-organisms dangerous to people and to the environment.

3. CONCLUSIONS

The adaptability and suitability of existing environmental regulations with respect to biotechnology will probably be a matter of further research. However, it seems that environmental regulations, as they stand now, were simply not designed to control the risks which could arise from accidental or deliberate release into the environment of new living organisms. Further, even if some aspects of these measures could be made applicable, there would still be significant areas of concern where various open environment releases would not be subject to these procedures.

In some instances it may be possible to extend present legislation to cover various aspects of environmental concern. Nevertheless, the number of regulations to be amended is relevant and the inherent complexity and importance of biotechnology application does suggest that this piecemeal approach will not provide complete, consistent and predictable regulation to ensure environmental protection and productive industrial growth.

11 Product-specific regulations related to biotechnology

Community regulations related to products, directly concerned with product safety and health protection, will be applied to some extent to products from biotechnology. Their description, however, does not fall within the scope of this study. Up to now, the only Commission proposal specifically related to biotechnology relates to pharmaceutical products. The most important consequences which have so far resulted from harmonization of pharmaceutical regulation within the Community have been the harmonization of the criteria of quality, safety and efficacy of drugs, the fact that tests and clinical trials performed in accordance with the Community rules no longer need to be repeated within the Community, harmonization of general requirements concerning labelling, mutual acceptance of tests on batches and the adoption of a common list of permitted colouring matters for use in medicines. In spite of this harmonization, differences in the decisions taken by the ten national authorities responsible for drug marketing are apparent. In order to reduce these differences, two committees have been set up: the Committee for Proprietary Medicinal Products and the Committee for Veterinary Medicinal Products. Member States can apply to these Committees to obtain advisory opinions on particular medicinal products.

As early as in 1967, and again in 1980, the Commission proposed the mutual recognition of national marketing authorizations for proprietary medicinal products for human use; this involves a single assessment but decentralized decisions. The Council, although not going so far, has, by adopting Directive 83/570/EEC and Recommendation 83/571/EEC, provided the national authorities, the Commission and the firms with tools for progressing towards these objectives.

Firms that have obtained prior marketing authorization in one Member State and request the forwarding of the application to at least two other Member States, can request the State to lodge their applications through the Committee. The Committee then issues an opinion, although it has no authority to oblige Member States to issue or refuse authorization.

In fact, the rules covering medicines (both human and veterinary) now exclude immunological, blood and radioactive products and homeopathy. However, immunological products are likely to be produced in the near future through biotechnology.

In October 1984, the Commission submitted to the Council five proposals for legislative measures for high technology medicinal products,[63] particularly those derived from biotechnology. On March 1985, the Economic and Social Committee expressed a favourable opinion on these proposals,[64] the European Parliament will deliver its opinion in the near future. The Council's final decision will probably be taken in 1986. The first proposal refers specifically to 'high technology and biotechnology' drugs, whereas the other four measures cover all medicines, but are considered to be specially useful for these 'high technology' drugs.

In the first proposal the competent authorities should be obliged to consult the Committee for Proprietary Medicinal Products and the Committee for Veterinary Medicinal Products before they decide to authorize, refuse or withdraw any high technology medicine. The firm concerned is a direct party to this procedure which runs parallel to the national examination procedure of which the deadline must be categorically adhered to. This consultation procedure should be initiated systematically in the case of biotechnological products, or at the request of the firms concerned, in the case of other high technology medicaments.

Furthermore, the Member States would be obliged to notify, before they adopt them, any draft national regulations affecting the manufacture, marketing or use of biotechnological medicines, so as to obviate further segmentation of domestic markets. This control is also related to the procedure for the provision of information in the field of technical standards and regulations laid down in Directive 83/189/EEC.[65]

12 Conclusions and recommendations

Environmental risks of primary concern are those associated with the intention or inadvertent release of new or exotic micro-organisms into the environment.

General guidelines can be worked out for biological and physical containment of micro-organisms to be used in large-scale industrial applications.

However, when deliberate release occurs, there is no need to distinguish between small- and large-scale applications because a quantity of a self-replicating micro-organism can never be small enough to ensure effective prevention of risks for the environment. Deliberate release of micro-organisms requires a previous case-by-case assessment of potential hazards.

Additional research in the field of microbial ecology and model ecosystems is needed to document further the safety or the potential hazards of deliberate release of new or exotic micro-organisms.

Additional experience and data are needed to be able to move from a case-by-case analysis to develop and implement standard data requirements, hazard assessment procedures and registration policies.

The adoption of an Ad-Hoc Directive is proposed (see Annex 12), intended to control risks from accidental and deliberate release of new and exotic micro-organisms where the general philosophy of preliminary exchange of information should be applied.

Notes

1. Bull, A. T., Holt, G., and Lilly, M. D., *Biotechnologies, International trends and perspectives*, Paris, Organization for Economic Co-operation and Development, 1982.
2. U.S. Office of Technology Assessment, 'Commercial Biotechnology: An International Analysis', Washington D.C., 1984.
3. Forecasting and Assessment in the Field of Science and Technology (FAST), *Europe 1995*, London, Butterworth Scientific, 1984.
4. U.S. Office of Technology, op. cit.
5. U.S. Department of Commerce-International Trade Administration, 'High Technology Industries: Profiles and Outlooks—Biotechnology', Washington D.C., 1984.
6. Op. cit.
7. Thomas, T., 'Chemistry and biology: An interface in oils', Leverhulme lecture, Annual Meeting Society of Chemistry and Industry, Liverpool, July 1982.
8. Bunch, T. D., Foote, W. C. and Spillet, J. J., *A new look on wild sheep, Bostid Developments*, Journal of the Board on Science and Technology for International Development, Office of International Affairs, National Research Council, Washington D.C., **4**, No. 3, October 1984.
9. World Health Organization (WHO) (1982), 'Health Impact of Biotechnology', Report on a WHO Working Group, Zurich, *Swiss Biotech*, **2**, No. 5, 1984.
10. 'Cornell Sees Market for Bovine GH', in McGraw-Hill's Biotechnology Newswatch, New York, **5**, No. 4, February 1985.
11. Sharp, W. R., Evans, D. A, Ammirato, P. V. DNAP Corp., 'Plant Genetics poised to revolutionize agriculture', *ECN Chemscope*, Biotechnology supplement, May 1984.

12. Ratledge, C., Oral presentation at the European Federation of Biotechnology Workshop on 'A Community Strategy for European Biotechnology', 1981.
13. Curtin, M. E., 'Microbial mining and metal recovery: Corporations take the long and cautious path', *Bio/technology*, May 1983.
14. Chakrabarty, A. M., 'Biodegradation and Detoxification of Environmental Pollutants', CRC Press, Boca Raton, Florida, 1982.
15. Simberloff, D., 'Community effects of introduced species', *Biotic crises in ecological and evolutionary time*, Academic Press, Chicago, Illinois, 1981.
16. Committee on Science and Technology, 'The environmental implications of genetic engineering', U.S. House of Representatives—Staff Report, 1984. Hearing of Clay, D. (Environmental Protection Agency).
17. Ibid., hearing of Alexander, M. (Cornell University).
18. Ibid., hearing of Sharples, F. (Oak Ridge National Laboratory).
19. Lesk, A. M., 'Protein structure and evolution: Similar amino acid sequences sometimes produce strikingly different three-dimensional structures', *Bio Essays*, ICSU Press by Cambridge University Press, 1985.
20. Giles, K. L. and Whitehead, C. M., 'Reassociation of a modified mycorrhiza with the host plant roots (Pinus Radiata) and the transfer of acetilene reduction activity', 48 *Plant and Soil*, **48**, 1977, pp 143–52.
21. National Institute of Health, 'Guidelines for research involving recombinant DNA molecules', Federal Register 48, Washington D.C., 1983, pp. 24556–81.
22. U.K. Guidelines, Advisory Committee on Generic Manipulation (ACGM), Guidance Notes 12, 13.
23. Mllewski, E., 'Large scale procedures under the Guidelines', *Recombinant DNA Technical Bulletin*, **5**, No. 2, June 1982.
24. European Federation of Biotechnology, 'Biotechnology in Europe', report to the FAST-Bio-Society Project of the Commission of the European Communities; Dechema, Frankfurt, 1983.
25. Sargeant, K. and Evans, C. G. T., 'Hazards involved in the industrial use of microorganisms', Study Contract

430/78/5, Commission of the European Communities, Brussels, 1978.
26. Evans, C., Preece, T. and Sargeant, K., 'Microbial Plant Pathogens: natural spread and possible risks in their industrial use', Study Contract XII/1059/81/EN of the Commission of the European Communities, Brussels, 1981.
27. European Federation of Biotechnology, Working party on Safety in Biotechnology Report, 'Safe Biotechnology', Dechema, Frankfurt, 1983.
28. Curtiss, R., 'Genetic manipulation of microorganisms: potential benefits and biohazards', Annual Review of Microbiology, **41**, 1976, pp. 507–33.
29. Stozky, G. and Babich H., 'Fate of Genetically-Engineered Microbes in Natural Environments', *Recombinant DNA Technical Bulletin*, **7**, No. 4, December 1984.
30. Anderson, J., 'Viability of, and transfer of a plasmid from *E.coli* K-12 in the human intestine', *Nature*, **225**, 1974, pp. 502–4. Levy, S. and Marshall, B., 'Risk assessment studies of *E.coli* host-vector systems', *Recombinant DNA Technical Bulletin*, **4**, pp. 91–7. Levy, S., Marshall, B. and Rowse-Eagel, D., 'Survival of *E.coli* host-vector systems in the mammalian intestine', *Science*, **209**, 1980, pp. 391–4. Smith, H., 'Survival of orally administered *E.coli* K-12 in alimentary tract of man', *Nature*, **255**, 1975, pp. 500–2.
31. Evans, C., Preece, T. and Sargent, K., op. cit.
32. Ibid.
33. Ibid.
34. European Federation of Biotechnology, 'Safe Biotechnology General Considerations', *Applied Microbiology and Biotechnology*, **21**, 1985, pp. 1–6.
35. Stozky, G. and Babich, H., op. cit.
36. Atkinson, B. and Sainter, P., 'Technological Forecasting for Downstream Processing in Biotechnology', Phase I, Intermediate Forecast Report, Series FAST, No. 6, EUR 8041 EN, Commission of the European Communities, Brussels, 1982.
37. Sargeant, K. and Evans, C. G. T., op.cit.
38. Ibid.
39. Rodger, G., 'Food Proteins—Current Trends and future developments', paper presented at Biotech '85 Conference (Europe), Online Publications, Pinner, U.K., 1985.

40. Ibid.
41. FSC Food Standards Committee Report on Novel Protein Foods, MAFF FSC/REP/62, HMSO, London, 1974.
42. 'Current Good Manufacturing Practices', Code of Federal Regulations, Title 21, Part 211.
43. Sharples, F. E., 'Spread of organisms with novel genotypes: Thoughts from an Ecological Perspective', *Recombinant DNA Technical Bulletin*, **6**, No. 2, June 1983.
44. Ibid.
45. Alexander, M., 'Microbial Ecology', John Wiley and Sons, New York, 1976.
46. Reanney, D., 'Extrachromosomal elements as possible agents of adaptation and development', Bacteriol Review, **40**, 1976, pp. 552–90.
47. Rissler, J. F., 'Research needs for Biotic Environmental Effects of Genetically-Engineered Microorganisms', *Recombinant-DNA Bulletin*, **7**, No. 1, Bethsda MD20205, USA, March 1984.
48. Stozky, G. and Babich, H., op. cit.
49. OECD AD-Hoc Group on Safety and Regulations in Biotechnology, Draft Report DSTI/85.23, Paris, 1985.
50. Milewski, E. and Tolin, S., 'Development of Guidelines for Field Testing of Plants Modified by Recombinant DNA Techniques', *Recombinant DNA Technical Bulletin*, **7**, No. 3, September 1984.
51. Sharples, F. E., op. cit.
52. Office of Technology Assessment (OTA), 'Impacts of Applied Genetics—microorganisms, plants and animals', Congress of the United States, Washington D.C., 1981.
53. Report to the President by an Interagency Task Force on Tropical Forests, 'The world's tropical forests: a Policy, a Strategy and Program for the United States', State Department, Publication No. 9117, Washington D.C., May 1980.
54. 'Engineering Herbicide Tolerance: When Is It Worth?', *Nature*, **2**, No. 11, November 1984.
55. Kidd, G., William Teweles and Co., 'The New Plant Genetics: Restructuring the Global Seed Industry', paper presented at Biotech '85 (Europe), Online Publications, Pinner, U.K., 1985.
56. Myers, N., 'Conversion of Tropical Moist Forests', report for

the Committee on Research Priorities in Tropical Biology of the National Research Council, National Academy of Science, Washington, DC., 1980.
57. Burley, W. and Courrier, K., 'A Genetic Cornucopia', Plant Breeding genetics and Bio-Technology Intelligence Report, Marketplace Intelligence Service, MN 55408, November 1984.
58. Waohtel, P., 'Seeds of Conflict', *International Wildlife*, November/December 1984.
59. OTA, op. cit.
60. Shattock, R., Janssen, B., Whitbread, R. and Shaw, D., 'An interpretation of the frequencies of host specific phenotypes of *Phytophtora infestans* in North Wales', Annales of Applied Biology, **86**, 1977.
61. OECD, op. cit.
62. Otaway, H., 'Regulation and Risk Analysis', unpublished report, 1985.
63. COM(84)437 and O.J. No. C293, Brussels, 5 November 1984.
64. E.S.C. (Economic and Social Committee) 317/85 and 329/85, Brussels, 1985.
65. Official Journal of the European Communities (O.J.), No. L109, 24 April 1983.

Annex 1

Insertion des biotechnologies (zone grisée) dans différents domaines (définition adoptée par FAST)

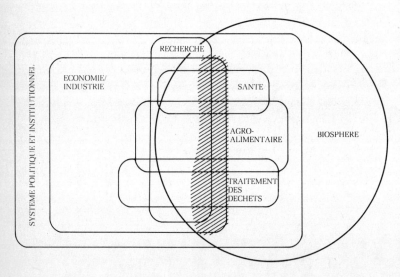

Source: Rapport FAST, EUROPE 1995, mutations technologiques & enjeux sociaux.

Annex 2

Some recent definitions of biotechnology

1. Biotechnologie, Eine Studie über Forschung und Entwicklung, Dechema (1976) (translation).
Biotechnology is concerned with the use of biological activities in the context of technical processes and industrial production. It involves the application of microbiology and biochemistry in conjunction with technical chemistry and process engineering.
2. Biotechnology—Report of a Joint Working Party, UK (1980).
The application of biological organisms, systems, or processes to manufacturing and service industries.
3. Biotechnology in Canada: Promises and Concerns (1980).
The application of biological organisms, systems, or processes to manufacturing or service industries.
4. Biotechnology: A Development Plan for Canada (1981).
The utilization of a biological process, be it microbial, plant or animal cells, or their constituents, to provide goods and services.
5. Biotechnology: a Dutch Perspective (1981).
The science of applied biological processes.
The science of the production processes based on the action of micro-organisms and their active components, and of production processes involving the use of cells and tissues from higher organisms (narrower definition). Medical technology, agriculture and traditional crop breeding are not generally regarded as biotechnology.
6. Biotechnology for Australia (1981).
The devising, optimizing and scaling-up of biochemical and

cellular processes for the industrial production of useful compounds and related applications. This definition envisages biotechnology as embracing all aspects of processes of which the central and most characteristic feature is the involvement of biological catalysts. Plant agronomy falls outside this definition but plants provide the raw material for most biotechnological processes, so research in plant breeding and productivity is of direct importance.

7. OTA Report—Impacts of Applied Genetics (1981).

The collection of industrial processes that involve the use of biological systems (in glossary).

The use of living organisms or their components in industrial processes.

8. FAST (Forecasting and Assessment for Science and Technology).

Sub-programme Bio-society—research activities.

The industrial processing of materials by micro-organisms and other biological agents to provide desirable goods and services.

9. European Federation of Biotechnology (1981).

The integrated use of biochemistry, microbiology and engineering sciences in order to achieve technological (industrial) application of the capabilities of micro-organisms, cultured tissue cells, and parts therefore.

10. IUPAC (International Unions of Pure and Applied Chemistry) (1981).

The application of biochemistry, biology, microbiology and chemical engineering to industrial processes and products (including here the products in health care, energy and agriculture) and on the environment.

Source: Bull, A. T., Holt, G. and Lilly, M. D., Biotechnology: International trends and perspectives (Paris: Organisation for Economic Co-operation and Development, 1982).

Annex 3

Major Events in the Commercialization of Biotechnology

1973 First gene cloned.

1974 First expression of a gene cloned from a different species in bacteria.
Recombinant DNA (r-DNA) experiments first discussed in a public forum (Gordon Conference).

1975 U.S. guidelines for r-DNA research outlined (Asilomar Conference).
First hybridoma created.

1976 First firm to exploit r-DNA technology founded in the United States (Genentech).
Genentic Manipulation Advisory Group (U.K.) started in the United Kingdom.

1980 *Diamond v. Chakrabarty*—U.S. Supreme Court rules that micro-organisms can be patented under existing law.
Cohen/Boyer patent issued on the techniques for the construction of r-DNA.
United Kingdom targets biotechnology (Spinks' report).
Federal Republic of Germany targets biotechnology (Leistungsplan).
Initial public offering by Genentech sets Wall Street record for fastest price per share increase ($35 to $89 in 20 minutes).

1981 First monoclonal antibody diagnostic kits approved for use in the United States.
First automated gene synthesizer marketed.
Japan targets biotechnology (Ministry of International Trade and Technology declares 1981 'The Year of Biotechnology').
France targets biotechnology (Pelissolo report).
Hoescht/Massachusetts General Hospital agreement.
Initial public offering by Cetus sets Wall Street record for the largest amount of money raised in an initial public offering ($115 million).
Industrial Biotechnology Association founded.
DuPont commits $120 million for life sciences R&D.
Over 80 NBFs had been formed by the end of the year.

1982 First r-DNA animal vaccine (for colibacillosis) approved for use in Europe.
First r-DNA pharmaceutical product (human insulin) approved for use in the United States and the United Kingdom.
First R&D limited partnership formed for the funding of clinical trials.

1983 First plant gene expressed in a plant of a different species.
$500 million raised in U.S. public markets by NBFs.

Source: Office of Technology Assessment-Commercial Biotechnology, January 1984.

Annex 4

BIOTECHNOLOGY: ACCORDING TO INDUSTRIAL SECTORS

Sector	Activities
Chemicals: organic (bulk)	ethanol, acetone, butanol
	organic acids (citric, itaconic)
organic (fine)	enzymes
	perfumeries
	polymers (mainly polysaccharides)
inorganic	metal beneficiation, bioaccumulation and leaching (Cu, U)
Pharmaceuticals	antibiotics
	diagnostic agents (enzymes, antibodies)
	enzyme inhibitors
	steroids
	vaccines
Energy	ethanol (gasohol)
	methane (biogas)
	biomass
Food	dairy, fish and meat products
	beverages (alcoholic, tea and coffee)
	baker's yeast
	food additives (antioxidants, colours, flavours, stabilizers)
	novel foods

Sector	Activities
	mushroom production
	amino acids, vitamins
	starch products
	glucose and high fructose syrups
	functional modifications of proteins, pectins
	toxin removal
Agriculture	animal feedstuffs
	veterinary vaccines
	ensilage and composting processes
	microbial pesticides
	Rhizobium and other N-fixing bacterial inoculants
	mycorrhizal inoculants
	plant cell and tissue culture (vegetative propagation, embryo productions, genetic improvement)
Service Industries	water purification
	effluent treatment
	waste management
	oil recovery
	analytical tools

BIOTECHNOLOGY: BASED ON VOLUME AND VALUE

Category	Activities
High volume, low value	methane, ethanol
	biomass
	animal feed
	water purification, effluent and waste treatment
High volume, intermediate value	amino and organic acids
	food products
	baker's yeast
	acetone, butanol
	polymers
	metals

Low volume, high value   antibiotics and other health care
                              products
                          enzymes
                          vitamins

Source: Bull, A., Holt, G., and Lilly, M., Biotechnology: International trends and perspectives (Paris: Organisation for Economic Co-operation and Development, 1982).

Annex 5

Some health care products for manufacture by r-DNA Technologies

Products	Potential uses/ Developments
Antibiotics	Infectious disease treatment
Vaccines	Immunization for disease prevention
Hepatitis A and B	
Rabies	
Influenza	
Malaria	
Diphtheria	
Polio	
Biological response modifiers	
Hormones	
Insulin	Treatment of diabetes
Human growth hormone	Treatment of dwarfism and acceleration of wound healing
B-endorphins	Treatment of pain
Thymosin alpha-1	Immune system stimulant
Erythropoietin	Stimulation of red blood cell production
Secretin	Treatment of digestive problems
Somatostatin	Treatment of pituitary disease
Interferons	Treatment of viral diseases and cancer

Biologicals
  Albumin   Blood expander
  Enzymes
    Urokinase   Dissolving blood clots
    Clotting factors VIII IX   Treatment of hemophilia
    Bradykinin   Treatment of high blood
                                pressure
    Glucosidase   Enzyme replacement therapy
                                (that is, Gauche's Disease)

Source: U.S. Dept. of Commerce, 1985.

Annex 6

Potential Chemical Products using Recombinant DNA Technologies

Products	Potential Uses
Commodity chemicals	Intermediates and bulk chemicals
Amino-acids	
Acetic acid	
Acetone	
Acrylic acid	
Adiptic acid	
Citric acid	
Ethanol	
Ethylene glycol	
Ethylene oxide	
Glycerol	
Methanol	
Propylene oxide	
Salicylic acid	
Industrial enzymes	
Oxidases	Production of glycols
Amylases	Production of sugars
Lignases	Production of phenols from wood
	cellulose for plastics

Cellulases	Production of fermentable sugars from wood cellulose

Fibers and plastics
   Cellululosics
   Silk
   Polyhydroxybutyrate
   Pullalans
	For use in textiles and plastics

Specialty chemicals	Manufacturing processes
Oils	Lubricants

Source: U.S. Dept. of Commerce: 'High technology Industries: Biotechnology—Profiles and Outlooks'. July 1984.

Annex 7

Food and Beverage Industries: Present Practice and Future Development

Industry	Comments
Brewing	Largest bulk biotechnology industry (700 million hectolitres per year). Materials used include malted barley, roasted barley, maize, rice, sorghum and wheat as components in fermentation wort. Flavour with hops and fermentation with *Saccharamoyces sp.* Extensive use of enzyme (starch and protein degrading) in production and storage. Some national industries maintain traditional methods; some work on new yeast strains (produced by hybridization and protoplast fusion) to utilize an increased range of substrates for fermentation (e.g. dextrins). Problems have been encountered with flavour components (e.g. 4-vinyl guaiacol). Scope for genetically engineered strains avoiding these flavour complications.
Milk fermentation	Wide variety of products including cheese, yoghurt, buttermilk, kefir and koumis, some of local or national origin. Large-scale production of cheese and yoghurt involves specific starter cultures. Largest problem is susceptibility of organisms concerned (*Streptococcus sp.*, *Labtobacillus sp.*) to lysis by

bacteriophage. Defined and single-organism cultures used in US and Australia but less favoured in Europe. Large scale for improvements in rate of fermentation and stability by genetic engineering. Microbial rennins have potential to replace animal products in production of curd (gelling of casein components of milk). Prospects for better process control and continuous processing.

Flavour component manufacture

Monosodium glutamate production. Other flavour-enhancing components include guanosine 5'-monophosphate, inosine, 5'-monophosphate and xanthosine 5'-monophosphate, which are produced commercially by *Corynebacterium glutamicium* and *Brevibacterium ammoniogenes*. Potential for strain improvement and alternative fermentation feedstocks.

Most colouring flavour components derived from plant sources. Potential exists for producing these (e.g. the «hot» principle of *Capsicum frutescens*) in plant cell cultures or in genetically engineered bacteria (long term).

Product	Process	Comments
Glycerol	*S. cerevisiae* «steered» fermentation	Possibility of fermentation in future without requirement for steering agents.
Acetic acid	Oxidation of ethanol by acetic acid bacteria	Biological system currently only for vinegar production. Commercial chemical production by methanol carbonylation estimated at 1.4 million t/y (US$ 500 million/y). Acetic acid is important chemical and fermentation feedstock. Future

Product	Process	Comments
		potential includes the conversion of CO_2/H_2 to acetic acid by *Acetobacterium woodii* and *Clostridium aceticum* and conversion of cellulose by thermophilic bacteria. Genetics of producer organisms poorly understood, restricting developments.
Citric acid	Produced from molasses by *Aspergillus niger*	Widely used food ingredient. Surface and submerged batch cultures both used currently although continuous process more efficient. World sales 0.2 million t/y (US$ 260 million/y). Molasses increasingly expensive as fermentation feedstock; starch (corn or wheat) under investigation as alternative. Possible use of genetic engineering to enable A. niger to degrade cellulose, etc. Yeast process using alkanes not now economical.

Product	Comments
Curdlan	Beta-1 → 3 glucan produced by *Alcaligenes faecalis* in high yield. Proposed use as gelling agent in food industry.

Alginates	Copolymer of DL-mannuronic and L-guluronic acids. Mainly produced from sea weeds but potential for development of bacterial e.g. from *Azotobacter vinelandii*. Many applications in food industry.
Xanthan	Substituted cellulose polymer from *Xanthanomonas campestris*. Widely accepted for use in food (e.g. forms stable gel in frozen and tinned foods, instant puddings, toppings) and for nonfood use with oil-drilling muds, as a paint stabilizer, flocculant for water clarification, and in gelled detergents.

Source: W.H.O. Working group on Health Impact of Biotechnology – Report – Swiss Biotech 2 (1984).

Annex 8

Waste Products Usable as Chemical/Fermentation Feedstocks and for Energy Production

Process industry and municipal wastes	Agriculture	Forestry
Molasses	Straw, bagasse	Sulfite liquor
Distillery waste	Coffee, cocoa	Bark, sawdust
Milk whey	coconut wastes	cellulose fibre
Solid and liquid waste from food industry	Fruit peels and waste	
Fishery waste	Tea waste	
Meat byproducts, abbatoir waste	Oilseed cake	
Sewage, garbage	Cotton waste	
Paper	Bran	
	Animal effluents	

Source: Smith, J. E.: Biotechnology. London, Edward Arnold, 1981 (Studies in Biology, No. 136).

Annex 9

Proposed EFB classification of micro-organisms according to pathogenicity

EFB Class 1

This class contains those micro-organisms that have never been identified as causative agents of disease in man and that offer no threat to the environment. They are not listed in higher classes or in Group E.

EFB Class 2

This class contains those micro-organisms that may cause disease in man and which might therefore offer a hazard to laboratory workers. They are unlikely to spread in the environment. Prophylactics are available and treatment is effective.

EFB Class 3

This class contains those micro-organisms that offer a severe threat to the health of laboratory workers but a comparatively small risk to the population at large. Prophylactics are available and treatment is effective.

EFB Class 4

This class contains those micro-organisms that cause severe illness in man and offer a serious hazard to laboratory workers and to people at large. In general effective prophylactics are not available and no effective treatment is known.

EFB Group E (environment risk)

This group contains micro-organisms that offer a more severe threat to the environment than to man. They may be responsible for heavy economic losses. National and international lists and regulations concerning these micro-organisms are already in existence in contexts other than biotechnology (e.g., for phytosanitary purposes).

These EFB descriptions differ slightly from those in existing European and American classifications. The proposed and existing classifications are compared in the Annex.

Annex 10

EXAMPLES OF CONJECTURAL RISK SCENARIOS

Energy: such applications often involve the direct release of genetically altered micro-organisms to the environment. Those used in Enhanced Oil Recovery are potentially serious human pathogens.

Mineral leaching: the large-scale use of micro-organisms may result in natural selection of strains that are infectious to human. The leaching may also enhance the generation of sulphuric acid which could cause serious acidification of fresh water sources.

Metal concentration from settling ponds or dilute water stream: the use of bacteria could transform some of the metals (for example, mercury) into organometallic compounds that are toxic to higher forms of life and could enter the food chain in the environment.

Oil-degrading bacteria to clean up oil spills might continue to consume oil after the spill is cleaned up, perhaps destroying oil resources.

Waste treatment: heavy metal ions might be transformed by micro-organisms into organic derivatives that are toxic to aquatic animals that take them up from the sediments. In other applications there is a potential public health threat from infectious bacteria being spread through aerosols generated by sewage treatment plants (that is, from air bubbled through activated sludge and sewage water splashing over rocks in trickling filter beds). These risks, already existing with conventional treatment facilities, could be increased by the use of genetically modified micro-organisms.

Lignin and cellulose degraders might attack living trees.

Agriculture: bacteria which prevent ice-nucleation might migrate to northern crops that require a freezing period to grow and might adversely alter cultivation. Sewage and other forms of waste water might be treated by genetically altered microorganisms and applied as fertilizers to the soil. This could result in harmful aerosols and groundwater contamination. Genetically engineered species could transfer genetic material to other plants, perhaps resulting in more vigorous weeds (and the need to use more herbicides), increased denitrification, increased crop disease susceptibility and changes in the niches and pathogenicities of plant viruses and soil bacteria. Nitrogen-fixing *Rhizobium* species, once able to infect monocotyledonous plants, could infect other species other than cultivated cereal crops, thus conferring advantage to weeds which could disrupt into many balanced ecosystems.

But see p 60

Annex 11

Genetic Research Timetable for Key Crops

		Identification, Duplication and Modification of Agriculturally Important genes	Routine growth of plant tissue in Laboratory culture conditions	Growth of first genetically transformed whole plant	First Plants altered by new technology available to breeder for commercial production	Growth of transformed plants on a routine basis
Major Cereals	Corn	now (zein, early maturity genes)	now	early 1990s	now	mid-1990s
	Wheat	1985–1987	now	early 1990s	1984–1986	mid-1990s
	Rice	1985–1987	now	late 1980s	now	early 1990s
	Barley	now (hordein, powdery mildew resistance genes)	now	1986–1988	1985–1987	early 1990s
	Sorghum	1987–1989	1984–1986	early 1990s	1988–1990	mid-1990s

		Identification, Duplication and Modification of Agriculturally Important genes	Routine growth of plant tissue in Laboratory culture conditions	Growth of first genetically transformed whole plant	First Plants altered by new technology available to breeder for commercial production	Growth of transformed plants on a routine basis
Oil Seeds						
	Oil Palm	now (nitrogen fixation genes)	now	early 1990s	1988-1990	mid-1990s
	Sunflower	1988-1990	now	late 1990s	now	after 2000
	Oilseed Rape	1985-1987	1984-1986	now	1984-1986	1986-1988
Forages		1984-1986	now	late 1980s	now	early 1990s
	Alfalfa	1986-1988	now	1985-1987	now	early 1990s
	Red Clover	now (nitrogen fixation genes)	now	early 1990s	now	mid-1990s
Vegetables						
	Tomatoes	1984-1986	now	1983-1985	1983-1985	1986-1988
	Lettuce	1985-1987	now	late 1980s	1985-1987	early 1990s
	Cucumber	1986-1988	1983-1985	mid-1990s	1984-1986	late 1990s
	Onion	1986-1988	1984-1986	early 1990s	now	mid-1990s
	Potato	now	now	1983-1985	now	1986-1988
	Carrot	1983-1985	now	1983-1985	1985-1987	1986-1988
	Beans	now (phaseolin)	1984-1986	1986-1988	1985-1987	early 1990s
	Peas	now (vicilin, legumin)	1984-1986	mid-1990s	now	late 1990s
	Brassicas	1983-1985	now	late 1980s		early 1990s
Grasses						
	Kentucky Bluegrass	late 1980s	1985-1987	mid-1990s	1986-1988	late 1990s
	Orchardgrass	late 1980s	1985-1987	mid-1990s	1986-1988	late 1990s

Wood, Plants					
	Fruit, Nut and ornamental trees	mid-1990s	1986–1988	late 1990s	early 1990s	after 2000
	Forest trees	mid-1990s	now	late 1990s	early 1990s	after 2000
Speciality Crops	Sugarbeets	1985–1987	now	early 1990s	1987–1989	mid-1990s
	Sugarcane	1987–1989	now	early 1990s	now	mid-1990s
	Cotton	1985–1987	now	early 1990s	1983–1985	mid-1990s
	Tobacco	now	now	1983–1985	now	1986–1988

Source: L. William Teweles & Co.

Annex 12

Main elements for a Directive on the notification of activities with biological agents.

A. Justifications:

Among the classical principles for the justification of the Directive the following aspects should be underlined:

- the use of biological agents in industrial activities and in the environment requires particular attention in order to control hazards for human health and the environment.
- such applications should be subject to *a priori* study by the manufacturer or importer and a notification to the competent authorities conveying mandatory information.

B. Main elements:

1. *Scope*: The definition of the scope shall cover all activities in which biological agents might be used and all products which might contain biological agents. Exempted from notification will be those activities and products already subject to notification or to approval procedures at European level concerned with protection of man and the environment. Also, biological agents considered safe will be exempted from notification.

Biological agents, defined as multicellular, unicellular,

subcellular microbial organisms, will be listed into three risk classes according to the potential hazard for man and the environment. Lack of knowledge concerning potential hazards will be a parameter for stricter classification.

2. *General provisions*: biological agents covered by the directive shall be used in a contained environment. Three levels of containment corresponding to the three risk classes will be established.

3. *Notification procedures*: Persons wishing to undertake work with biological agents in quantities equal or above 10 liters shall submit a notification to the competent authorities. Information shall be submitted with respect to:

- the identity of the biological agent and the quantities involved;
- the proposed classification and the proposed level of the containment;
- the foreseeable risks, immediate or delayed, for men and the environment;
- the geographical location of the activity and the main environmental characterisics;

4. *Other provisions*: Specific provisions will relate to the:

- duties of the competent authorities;
- follow-up information;
- procedures for down-grading containment levels;
- exchange of information;
- confidentiality;
- adaptation of the Annexes in line with technical progress.

Index

agricultural applications 22–5, 140
  crop breeding goals 69
  risks from 5–7, 59–81, 154
  see also plant cultivation
allergies 42, 56
animal husbandry 22–3
  growth hormones 23
  infection by pesticides 59
anther and pollen culture 66
antibiotic resistant traits 3, 44
antibodies 139
  produced by cell fusion 13
atrazine 73–4

Bacillus subtilis 48
Bacillus thuringiensis 24, 74–5
Bates (1956), cited 68
Belgium, regulatory measures in 8, 83–4, 98
bioelectronics 20–1
biogas 25–6, 27, 139
biological diversity 76–9
biomedical products 57–8, 142–3
biotechnology
  definitions of 10–11, 135–6
  major events in commercialization of 137–8
  products of 14–15, 139–45
bovine growth hormone (BGH) 23
brewing, 21, 146
bulk chemicals 18–19, 139

cell fusion 57, 66, 80
  described 12–13
cellular biology 11
Chakrabarty, Dr A. 29, 137
chemical biotechnology products 18–20, 139, 144

Concertation Unit for Biotechnology in Europe (CUBE) 10, 11, 102
containment problems 39, 40–1, 47, 49–50
Current Good Manufacturing Practices (US) 5, 57, 58

Denmark, regulatory measures in 7–8, 84–6, 98
detergents 20
diagnostic reagents 42–3, 50, 139

ecosystems, disturbance of 1, 6, 7, 31–7, 39, 59, 60–2, 63, 65, 154
energy production 25–7, 59, 139, 153
  biomass production and conversion 25
engineering research 13
environmental protection 1, 6, 11, 28–9, 44–51, 59–81
  existing regulatory framework for 105–23
  regulatory guidelines/measures 2, 7–9, 39, 67–8, 83, 85, 88, 89, 90, 92–3, 95, 97, 98
  risk scenarios 153–4
  risks from accidental release 114–23, 128
  risks from environmental use 105–13, 128
enzymes 19, 21, 22, 42, 72, 139, 141, 143
  novel enzymes 54
Escherichia coli (*E.coli*) 3, 21, 30, 43–4, 48, 64
European Economic Community Biotechnology Regulations Interservice Committee (BRIC) 102

European Economic Community (cont.)
   Biotechnology Steering Committee (BSC) 101, 102
   Community regulatory measures 99–104
   Council Directives: 67/546/EEC 106, 112; 75/442/EEC 120–1; 76/464/EEC 122–3; 76/768/EEC 112–13; 78/319/EEC 121; 79/831/EEC (6th Amendment to 67/546/EEC) 106–12; 82/501/EEC 114–17; 83/189/EEC 127; 83/570/EEC 126; 83/571/EEC 126; 84/360/EEC 123; 84/631/EEC 122; 85/337/EEC 118–20
   member countries' regulatory systems 7–8, 82–98; *see entry for individual country*
   Recommendation 82/472/EEC 83, 84, 90, 93, 101
European Federation of Biotechnology 49, 136
   proposed micro-organism classification 151–2

feedstocks 18, 140
fermentation processes 14. 18, 41, 51–2
fibres and plastics 19, 145
flavourings and additives 21, 53, 139, 147
food processing 21–2, 139, 146–9
Forecasting and Assessment of Science and Technology (FAST) report 11, 100, 103, 136
France, regulatory measures in 8, 89–90, 98

Gasohol programme 25, 26, 139
genetic manipulation 20, 38, 73, 76
genetic recombination 12
   see also recombinant r-DNA technology
genetic transfer 4, 6, 63–6, 67, 71–3, 75
genotype variation 51–2, 58
German Health Authorities 49
Germany, Federal Republc of
   regulatory measures in 8, 87–8, 98
Giles, K. L. 36
Greece, regulatory measures in 8, 90, 98

Health and Safety Executives (UK) 49, 96

health care 17–18, 142–3
herbicide resistance 73–4
hormones 142
hydrocarbon pollution 28, 153

industrial processing applications 38–58
insecticides *see* pest control, pesticides
interferons 57, 142
interspecific hybridization 66
Ireland, regulatory measures in 8, 91, 98
Italy, regulatory measures in 8, 92–3, 98

Japan, r-DNA work in 100, 138

Kohler and Milstein 13

laboratory design and practice 39
legislation *see under individual country and* European Economic Community
Luxembourg, regulatory measures in 8, 93, 98

major accident hazards 114–17
metal extraction 27, 28
metal leaching and concentration 62, 153
metal recovery 27–8
micro-organisms
   accidental release of 47–8, 116–17, 128, 153
   EEC regulatory measures on 105, 112, 116–17, 120
   EFB classification of 49–50, 151–2
   genetic stability of 63–5
   in the environment 62–6, 128, 153
milk fermentation 21, 146
milk production 23
mineral leaching 153
molecular biology 11, 70
monoclonal antibodies 13, 20, 23, 138
mutation, in micro-organisms 51–3
myco-protein 22
mycorrhizal fungi 72, 140
Myers, Norman (*Conversion of Tropical Moist Forests*) 77

National Institute for Biological Standards and Control (NIBSC) 57

Netherlands 100; regulatory measures in 8, 93–5, 98
Netherlands Microbiological Society 49
nitrogen fixation genes 24, 25, 36, 71–2, 73, 140, 154
nodulation genes 71
novel enzymes 54
novel food protein 53–7
  guidelines to assessing hazards of 55–6
novel foods 139

oils and fats 20, 145
oil recovery 26, 62, 140, 153
oil spills 153
Organization for Economic Cooperation and Development (OECD) 10, 91, 101
  Safety and Regulation questionnaire 84, 85, 88–95 *passim*, 97

Peronospora tabacina 45
pest control 23–4, 62
pesticides 23–4, 65, 140
  pollution from 28
pest-resistant plants 74
pharmaceutical products 17–18, 125, 138, 139
phenotype variation 51, 52–3, 58
plant cultivation 6–7, 24–5, 66–81
  biological diversity 76–9
  biotechnological risks in 45–7, 67
  clonal propagation 25
  crop variability and vulnerability 78, 79–81
  herbicide and pest resistance 73–5
  macrominerals and 71–3
  nitrogen fixation 24, 25, 36, 71–2, 73
  research in 66–9
  research timetable for key crops 155–7
  resistance to environmental stress 69–71
  seed industry 75–6
plasmids 43–4, 63–5, 101
plastics 19, 20, 144
pollution control 26, 27, 28–9, 50–1, 62, 122–3
processing and production in biotechnology
  advantages over chemical processing 15

  disadvantages 15–16
  risks in 1, 3–5, 31, 40–51
  techniques of 12–16
protein engineering 19
  see also genetic manipulation, genetic transfer
proteins 22
  see also novel food proteins
protoplast fusion *see* cell fusion
Pseudomonas syringae 25, 29
Puccinia striiformis 46

Recombinant DNA (r-DNA) technology 12, 17, 30, 57, 63, 67, 80, 100, 101, 116, 137–8
  compliance with EEC recommendations on 83, 84, 90, 93
  EEC concern with 101–2
  guidelines and regulations on 7–8, 39, 67–8, 85, 87, 93–4, 95–6, 98, 137
  risks arising from 38–40, 43–4
  uses of 12, 13, 19, 144–5
regulations 2, 7–9
  see under individual country and European Economic Community
research
  risks in 1, 2–3, 38–40
  safety guidelines 30
Rhizobium 25, 71, 72, 140
risks
  from agricultural applications 5–7, 59–81, 154
  from industrial applications 38–58
  in processing and production 1, 3–4, 31, 40–51
  in research 2, 2–3, 38–40

safe work practices 4, 50–1
Sharples, F. E. (*Spread of Organisms with Novel Genotypes*) 62
somaclonal variation 66
speciality chemicals 19–20, 139
Sweden, r-DNA work in 100

Thiobacillus bacteria 27
tobacco blue mould 45

United Kingdom
  Department of Health and Social Security 49, 96

United Kingdom (cont.)
  guidelines for r-DNA experiments 39, 93
  Health and Safety Executives 49, 96
  regulatory measures in 7–8, 95–7, 98
United Nations FAO/WHO Expert Committee on Food Additives 21
United States
  Department of Health and Human Services 49
  Food and Drug Administration 57
  National Institutes of Health 67; guidelines on r-DNA experiments 39, 67–8, 83, 90, 91
  Office of Technology Assessment (OTA) report 10, 11, 19, 87, 136

United States gasohol programme 25, 26

Vaccines 42–3, 50, 57, 138, 139, 140, 142
Vavilov centres 77

Waste, treatment of 26–7, 29, 48–9, 120–2, 140, 150, 153
Whitehead, H. C. H. 36
working environment 41–4
World Health Organization 49, 57

Yellow rust 46

Zea diploperennis 78